AF904811

Pharmaceutically Relevant Microbial Enzymes

Vivek Sharma • Richa Salwan
Editors

Pharmaceutically Relevant Microbial Enzymes

Insights into Enzymatic Diagnosis and Product Development

Springer

Editors
Vivek Sharma
University Centre for Research and Development
Chandigarh University
Mohali, Punjab, India

Richa Salwan
Department of Basic Sciences
College of Horticulture and Forestry
Hamirpur, Himachal Pradesh, India

ISBN 978-981-95-4940-5 ISBN 978-981-95-4941-2 (eBook)
https://doi.org/10.1007/978-981-95-4941-2

© The Editor(s) (if applicable) and The Author(s), under exclusive license to Springer Nature Singapore Pte Ltd. 2026

This work is subject to copyright. All rights are solely and exclusively licensed by the Publisher, whether the whole or part of the material is concerned, specifically the rights of translation, reprinting, reuse of illustrations, recitation, broadcasting, reproduction on microfilms or in any other physical way, and transmission or information storage and retrieval, electronic adaptation, computer software, or by similar or dissimilar methodology now known or hereafter developed.
The use of general descriptive names, registered names, trademarks, service marks, etc. in this publication does not imply, even in the absence of a specific statement, that such names are exempt from the relevant protective laws and regulations and therefore free for general use.
The publisher, the authors and the editors are safe to assume that the advice and information in this book are believed to be true and accurate at the date of publication. Neither the publisher nor the authors or the editors give a warranty, expressed or implied, with respect to the material contained herein or for any errors or omissions that may have been made. The publisher remains neutral with regard to jurisdictional claims in published maps and institutional affiliations.

This Springer imprint is published by the registered company Springer Nature Singapore Pte Ltd.
The registered company address is: 152 Beach Road, #21-01/04 Gateway East, Singapore 189721, Singapore

If disposing of this product, please recycle the paper.

We dedicate this book to the countless scientists, researchers, and innovators who have devoted their lives to unlocking the secrets of the microbial world. Their tireless efforts and unwavering curiosity have paved the way for the groundbreaking advancements in medicine that we now take for granted. This work is a tribute to their collective contributions and an acknowledgment of the profound impact they have had on human health. To the families who supported us throughout this journey, we extend our deepest gratitude. Your patience, encouragement, and understanding provided the foundation upon which this project was built. Your support in every form—from late-night conversations to quiet moments of encouragement—made this work possible.

We are immensely grateful to the esteemed contributing authors for their invaluable expertise and dedication. Your insights have enriched this volume beyond measure, transforming it from a mere collection of ideas into a comprehensive and authoritative resource. Your commitment to sharing knowledge and your passion for your respective fields are a true inspiration. Finally, we dedicate this book to our students, the future leaders of science and medicine. May the knowledge and ideas presented here ignite your curiosity and fuel your passion to continue the pursuit of scientific discovery. We hope this book serves as a starting point for your own transformative journeys in the world of microbial enzymes and their boundless potential.

Dr. Vivek Sharma
Dr. Richa Salwan

Foreword

Microbial enzymes have long been at the forefront of biotechnology and pharmaceutical innovation, offering sustainable and efficient alternatives to conventional chemical synthesis. The book, *Pharmaceutically Relevant Microbial Enzymes*, brings together a collection of scholarly contributions that explore the scientific, industrial, and therapeutic significance of microbial enzymes, emphasizing their transformative role in the pharmaceutical sector. This volume provides a well-structured and comprehensive overview of microbial enzyme research—from fundamental concepts to advanced therapeutic applications.

Chapter 1 sets the stage by introducing microbial enzymes, highlighting their current status, technological advancements, and future prospects in the pharmaceutical industry. Chapter 2 discusses a critical dimension often overlooked—the risk assessment and adherence to international biosafety guidelines for enzyme-based therapeutics, particularly those used in enzyme replacement therapy. The succeeding chapters explore individual enzymes of major pharmaceutical relevance. Chapter 3 presents microbial uricase as a potent therapeutic agent for managing hyperuricemia and related disorders. Chapter 4 delves into the evolution and pharmaceutical applications of streptokinase, a life-saving thrombolytic enzyme widely used in cardiovascular therapy. Chapter 5 examines the therapeutic potential of nattokinase and collagenases, focusing on their roles in fibrinolysis and tissue regeneration. Further, Chap. 6 provides an insightful overview of microbial sacrosidase, elaborating on its mechanism of action and applications in digestive enzyme therapy. Chapter 7 focuses on microbial lactase, emphasizing its role in alleviating lactose intolerance and improving gastrointestinal health. The book then advances to Chap. 8, which explores the structure, catalytic properties, and medical significance of microbial glutaminases in cancer therapy and metabolic regulation. Continuing the theme of therapeutic enzymes, Chap. 9 highlights L-asparaginase, one of the most widely utilized microbial enzymes in leukemia treatment, while Chap. 10 discusses microbial cholesterol oxidase, its structural and functional attributes, and emerging applications in clinical diagnostics and drug design. Finally, Chap. 11 concludes with a futuristic perspective on enzyme immobilization, presenting its role in enhancing enzyme stability, reusability, and industrial scalability for pharmaceutical applications.

Altogether, this book provides a holistic view of the field—linking molecular insights with biotechnological innovations and real-world medical applications. It serves as a valuable resource for students, researchers, and professionals in microbiology, biotechnology, and pharmaceutical sciences. Each chapter reflects the growing synergy between microbial biocatalysis and modern drug development, underscoring the immense potential of microbial enzymes in addressing global healthcare challenges.

I am confident that *Pharmaceutically Relevant Microbial Enzymes* will inspire readers to explore new dimensions of enzyme research and catalyze further innovations that bridge microbial diversity with pharmaceutical advancement.

College of Horticulture & Forestry
Neri, Hamirpur, Himachal Pradesh, India
27 October 2025

Dharam Pal Sharma,

Preface

The discovery and harnessing of microbial enzymes represent one of the most transformative chapters in modern pharmacology and medicine. For centuries, the pharmaceutical industry relied on complex chemical syntheses to produce therapeutic agents, a process often fraught with environmental concerns and limitations in scalability. The modern era is witnessing a paradigm shift, where the intricate and elegant catalytic machinery of microorganisms is being leveraged to create a new generation of diagnostic tools and therapeutic products. This book, *Pharmaceutically Relevant Microbial Enzymes*, is an exploration of this exciting frontier. Our goal is to provide a comprehensive, interdisciplinary perspective that bridges the foundational principles of microbial enzymology with their direct and impactful applications in human health. With a focus on the exploration of microbial enzymes for various pharmaceutical applications, this book caters the diverse audience, including undergraduate and graduate students, researchers, and professionals in the fields of microbiology, biotechnology, pharmacology, and clinical diagnostics.

The book delves into the pivotal role of microbial enzymes in diagnostics. We uncover how the exquisite specificity and efficiency of these biocatalysts enable precise, rapid, and often noninvasive detection of biomarkers. From well-established clinical assays for glucose and cholesterol to novel applications in infectious disease and genetic testing, we highlight the diagnostic power of enzymes and the innovative research driving this field forward. For instance, glucose oxidase is a valuable enzyme for assessing glucose levels in biological solutions such as blood and urine. Further, the book also explores the role of microbial enzymes in biotransformation. We explore how microbial enzymes facilitate the synthesis of complex active pharmaceutical ingredients, including the production of highly specific chiral drugs and the modification of existing molecules to improve their efficacy or reduce side effects. This section also covers the exciting potential for developing enzyme-based therapeutics to directly treat diseases, such as L-asparaginase, which is well known for its excellent therapeutic responses in treating acute lymphoblastic leukemia.

This volume is the culmination of insights from leading experts who have dedicated their careers to unlocking the potential of microbial biocatalysts. We hope that the knowledge contained within these pages will not only serve as a valuable resource but also inspire the next generation of scientists to further innovate and explore the vast, untapped potential of the microbial world for the benefit of human health.

Mohali, Punjab, India Vivek Sharma
Hamirpur, Himachal Pradesh, India Richa Salwan

Acknowledgments

This book is a product of immense collaboration and dedication. It would not have been possible without the support and contributions of numerous individuals and institutions. We would like to express our sincerest thanks to all who were instrumental in the creation of this volume. First and foremost, we are deeply grateful to our esteemed contributing authors. Their profound knowledge, meticulous research, and commitment to excellence have shaped the very core of this book. We thank them for their tireless efforts in crafting their respective chapters, sharing their valuable insights, and contributing to the intellectual richness of this work. Their dedication has been the very backbone of this project, and we are honored to have collaborated with such a distinguished group of experts.

Both the editors wish to express our profound appreciation to our colleagues at *Chandigarh University, Gharuan, Mohali, Punjab, India, and College of Horticulture and Forestry, Neri, Hamirpur, Dr YSP University of Horticulture and Forestry, Solan, India*. The supportive and intellectually stimulating environment provided by the university has been crucial to our work. We thank the administration for fostering a culture that encourages academic and research pursuits, and our peers for their valuable discussions and camaraderie.

A special acknowledgment is extended to our friends and family. Your unwavering belief in us, your patience during demanding times, and your constant encouragement provided the motivation and strength to see this project through to completion. The constant support and conversations has been a constant source of inspiration. Finally, we extend our gratitude to the publishing team for their expertise and guidance throughout the entire process, from manuscript preparation to final publication. Their professionalism and hard work have been essential in bringing this book to fruition.

Thank you to everyone who played a role, no matter how small, in bringing this project to life.

Vivek Sharma
Richa Salwan

Contents

Editors and Contributors

About the Editors

Vivek Sharma is currently working as Associate Professor, University Centre For Research and Development, Chandigarh University. He did his PhD at CSIR-IHBT Palampur. He is having demonstrated expertise on exploring microbial enzymes such as antifungal proteins, multicopper oxidase, proteases, and various other enzymes for industrial application. At present he is Associate Editor of *Chemical & Biological Technologies in Agriculture* (CBTA) and Academic Editor of *PLoS ONE*. He has edited three international books and authored one lab manual. He has successfully guided three PhD students. He also secured grant and successfully completed project as PI as well as Co-PI. He has published more than 85 articles, including research articles, reviews, and book chapters with citations of over 17,500 and an H-index of 26. He is a reviewer of more than 80 international journals.

Richa Salwan is currently working as Assistant Professor (Microbiology), Department of Basic Sciences, College of Horticulture and Forestry, Neri, Hamirpur (Dr YSP University of Horticulture and Forestry, Solan, India). She completed her PhD in Microbiology at Academy of Scientific and Innovative Research (AcSIR). Presently, she is Academic Editor of *PLoS ONE*. Her expertise involves exploration of extremophilic microorganism for industrially relevant enzymes and mining their genomes for enzymes and secondary metabolites. She has published several research papers in journals of international reputations. She has also edited three international books and authored one lab manual. She has published more than 80 research articles that include research articles, reviews, and book chapters with a total citation of over 2200 and an H-index of 21. She is a reviewer of several international journals.

Contributors

Natasha Sharma Awasthi University Centre for Research and Development, Chandigarh University, Mohali, Punjab, India

Tuyelee Das University Centre for Research and Development (UCRD), Chandigarh University, Mohali, Punjab, India

Mansour Ghorbanpour Department of Medicinal Plants, Faculty of Agriculture and Natural Resources, Arak University, Arak, Iran

Rachna Shashikant Jadhav Faculty of Applied Sciences and Biotechnology, Shoolini University, Solan, Himachal Pradesh, India

Randhir Kaur University Centre for Research and Development, Chandigarh University, Mohali, Punjab, India

Saurabh Kulshrestha Faculty of Applied Sciences and Biotechnology, Shoolini University, Solan, Himachal Pradesh, India

Ajay Kumar Department of Industrial Microbiology, Jacob Institute of Biotechnology and Bioengineering, Sam Higginbottom University of Agriculture, Technology and Sciences, Prayagraj, Uttar Pradesh, India

Fauziah Marpani Faculty of Chemical Engineering, Universiti Teknologi MARA, Shah Alam, Selangor, Malaysia

Ewa B. Moliszewska Institute of Environmental Engineering and Biotechnology, University of Opole, Opole, Poland

Ghazala Mustafa Department of Plant Sciences, Faculty of Biological Sciences, Quaid-i-Azam University, Islamabad, Pakistan

Nivedita Priya Department of Medical Laboratory and Life Science, Parul Institute of Allied Health Sciences (PIAHS), Faculty of Medicine, Parul University, Vadodara, Gujarat, India

Dhruvi Patel Parul Institute of Paramedical and Health Sciences (PIPHS), Parul University, Vadodara, Gujarat, India

Richa Salwan College of Horticulture and Forestry, Dr YS Parmar University of Horticulture & Forestry, Neri, Hamirpur, Himachal Pradesh, India

Himanshu Sharma Amity School of Biological Sciences, Amity University Punjab, Mohali, Punjab, India

Rhydum Sharma University Centre for Research and Development (UCRD), Chandigarh University, Mohali, Punjab, India

Sneh Sharma Department of Biotechnology, College of Horticulture and Forestry, Dr YS Parmar University of Horticulture and Forestry Solan, Hamirpur, Himachal Pradesh, India

Vivek Sharma Department of Biotechnology, College of Horticulture and Forestry (Dr YS Parmar University of Horticulture and Forestry Solan), Neri, Hamirpur, Himachal Pradesh, India

University Centre for Research and Development, Chandigarh University, Mohali, Punjab, India

Smriti Singh Department of Anaesthesia and Operation Theatre Technology, Chandigarh Pharmacy College, Chandigarh Group of Colleges Jhanjeri (Mohali), Sahibzada Ajit Singh Nagar, Punjab, India

Priyanka Sood University Centre for Research Development (UCRD), Chandigarh University, Mohali, Punjab, India

Arti Thakur Department of Allied Health Sciences, Desh Bhagat University, Gobindgarh, Punjab, India

Mahendra Kumar Varma Department of Paramedical Sciences, Vivekananda Global University, Jaipur, Rajesthan, India

1 Introduction to Microbial Enzymes: Present Status and Future Scope in Pharmaceutical Industry

Richa Salwan and Vivek Sharma

Abstract

Microbial enzymes have immense applications in the pharmaceutical industry, due to their advantages over chemical alternatives being environmentally friendly, high specificity and efficiency, and ability to function under extreme conditions. These enzymes are extensively utilized in the pharmaceutical sector for enzyme replacement therapy (ERT) and as alternative therapeutics for various conditions, including cancer, metabolic disorders, and inflammatory conditions. The applications of these enzymes include diagnosis as well as treatments such as lipases for chiral drug synthesis, proteases for anti-inflammatory agents, asparaginase for leukemia treatment. In addition to this, microbial enzymes also have been explored as thrombolytic and digestive agents. This chapter provides an overview of the current status and future scope of microbial enzymes in pharmaceutical applications while acknowledging challenges such as immunogenicity, tissue specificity, and others.

Keywords

Proteins · Therapeutics · Diagnostic · Treatment

R. Salwan
College of Horticulture and Forestry, Dr YS Parmar, University of Horticulture & Forestry, Neri, Hamirpur, Himachal Pradesh, India

V. Sharma (✉)
University Centre for Research and Development, Chandigarh University, Mohali, Punjab, India

© The Author(s), under exclusive license to Springer Nature Singapore Pte Ltd. 2026

V. Sharma, R. Salwan (eds.), *Pharmaceutically Relevant Microbial Enzymes*,
https://doi.org/10.1007/978-981-95-4941-2_1

1.1 Introduction

The utilization of enzymes in industrial applications began in the mid-nineteenth century with fungal enzymes, followed by the introduction of bacterial enzymes approximately two decades later (Haile and Ayele 2022). Approximately 60% of industrial enzymes are produced from fungi, 24% from bacteria, 4% from yeast, and the rest 10% from plant and animal sources (Fasim et al. 2020). The pharmaceutical sector is constantly seeking innovative and sustainable approaches for drug discovery, development, and manufacturing. Traditional chemical-based methods that involve the usage of harsh reagents cause side effects and generate significant waste leading to high production costs and environmental concerns (Sheldon and Woodley 2018). In contrast, biocatalysis, particularly the microbial enzymes, offers an attractive alternative. The enzymes from microbial sources due to their fast growth, specificity and efficiency to perform under extreme conditions are preferred (Bornscheuer 2002). The pharmaceutical industry has witnessed a surge in the adoption of enzyme technologies for research, development, and manufacturing.

The concept of therapeutic enzymes dated back to the 1960s, with the exploration of enzyme replacement therapies (ERT) for genetic deficiencies (Vellard 2003). In particular, the microbial enzymes have emerged as powerful tools for the synthesis of complex pharmaceutical molecules including chiral intermediates, due to their high enantio-selectivity. Further, the diversity of microorganisms of both bacterial and fungal origin is an immense untapped reservoir of novel enzymes with a wide range of catalytic activities (Adrio and Demain 2014). Furthermore, the advantages of using genetic engineering have played a crucial role transforming these microorganisms into "cell factories" for various enzyme production (Ferrer et al. 2016). Therefore, the microbial enzymes are of economic importance for various industrial applications including pharmaceutical sector. The enzymes of microbial sources are indispensable due to their usage in transforming the secondary metabolites with high enantio-selectivity (Lorenz and Nett 2018). The advancements in the field of biotechnology, genetic engineering, and protein engineering techniques accelerated the production and exploration of enzymes for treating various diseases (Vellard 2003). In addition, microbial enzymes have been employed as therapeutic agents or for diagnostic purposes (Rani and Singh 2017). The comparatively less side effects further enhance the relevance of these enzymes for pharmaceutical applications (Yadav and Yadav 2018).

This provides an overview of the microbial enzymes in the pharmaceutical industry along with their advantages and the challenges. Furthermore, the chapter also highlights the transformative potential of emerging technologies in unlocking new possibilities for drug manufacturing, personalized medicine, and sustainable pharmaceutical production (Hassan et al. 2024).

1.2 History of Microbial Enzyme Exploration in Pharmaceutical Sector

Enzymes particularly of microbial origin are of vital importance in different industrial sectors, due to their versatile nature and abilities to perform under extreme conditions. The market potential of these enzymes has expanded due to their eco-friendly nature, ability to perform under extreme conditions, and ease of genetic manipulation. The global enzyme market is set to reach a substantial $525 billion by 2030, growing at a 6.7% annual rate. Within this, microbial enzymes are increasingly vital for antitumor therapy, with L-asparaginase and L-glutaminase leading the way, holding 16.5% and 9.5% market shares, respectively. These enzymes have opened up new avenues for fighting cancers like leukemia, lymphosarcoma, and breast cancer, boosting their demand in pharmaceuticals. However, their widespread use faces challenges such as a short half-life, immunogenicity, and toxicity. The industrial applications, industrial production, purification, formulation, and commercial use of microbial enzymes have been reviewed by Hassan and his colleagues in 2024 (Hassan et al. 2024). It has been observed that keeping in view of the importance, the efforts on the enzyme technology for pharmaceutical sectors and other therapeutic interventions are still gaining the major attention (Fig. 1.1).

The exploration and utilization of microbial enzymes in the industrial sector represent a fascinating journey, evolving from empirical observations to sophisticated biotechnological interventions. The early twentieth century saw the beginnings of industrial enzyme use, mainly in the food and textile industries. For instance, fungal enzymes were incorporated into bread making to enhance dough quality. Similarly, microbial proteases from *Aspergillus oryzae* became popular in the 1910s and 1920s for textile desizing (Gupta et al. 2002). During this time, enzymes were typically

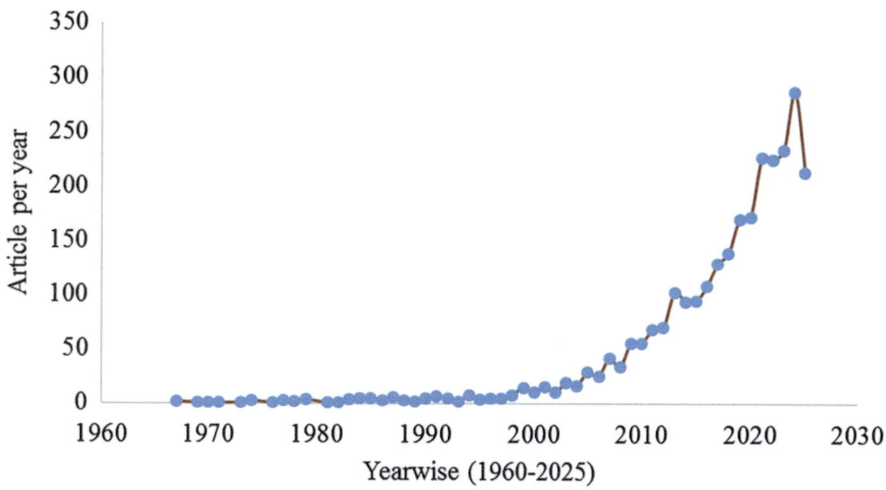

Fig. 1.1 Trends on microbial enzymes exploration in pharmaceutical sectors as per Scopus database dated 04-August-2025

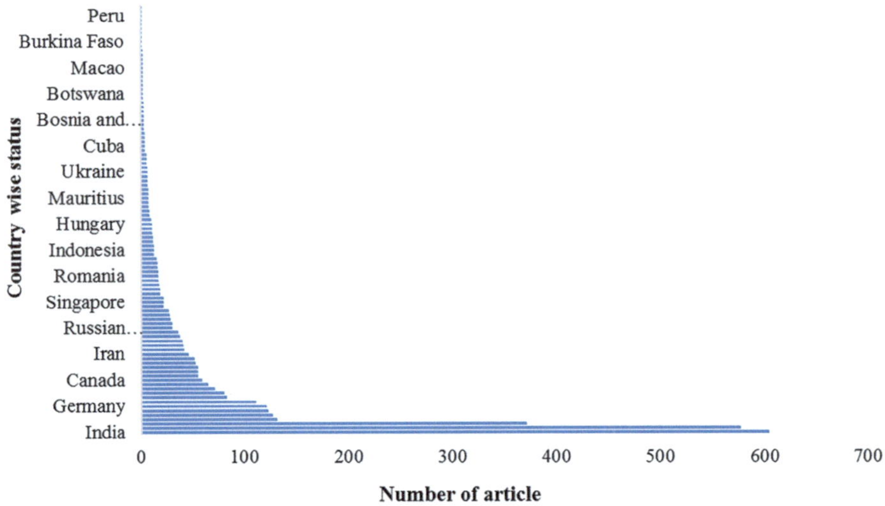

Fig. 1.2 Country-wise status of microbial enzymes exploration in pharmaceutical sectors as per Scopus database dated 04-August-2025

crude preparations from simple fermentation, and there was a limited understanding of their structure and function. The primary goal was large-scale production for industries focused on high-volume, lower-value products.

The mid-twentieth century marked a period of significant advancement, driven by breakthroughs in fermentation technology and in depth understanding of microbial physiology. The success of penicillin production during second World War attained the focus of microbial fermentation and microbial-based products (Demain and Sanchez 2009a, b). Countries like India, China, and the USA are leading the sectors (Fig. 1.2). For example, the role of amylases from *Bacillus* species in starch processing for glucose syrup production is a mile stone for the food industry (Crueger and Crueger 1990). The usage of proteases and lipases in detergent industry emerged as a key step in enzyme formulations (Godfrey and West 1996). Thereafter, efforts on characterization of microbial strains for enzyme production are more streamlined and much focused and targeted.

In the late twentieth century, a major transformative change was observed largely due to the developments in recombinant DNA technology. The developments in genetic engineering revolutionized the cloning of genes encoding enzymes from microbial hosts and subsequently their expression in heterologous host systems such as like *Escherichia coli*. The breakthrough in cloning in cloning technologies was able to address the concern for large-scale production of enzyme and purification issues (Cherry and Fidantsef 2003). The recombinant DNA technologies significantly addressed these issues and also opened the opened avenues for protein engineering, using site directed mutagenesis and codon optimization through which targeted amino acid residues can be used for improving the enzyme stability, activity, specificity, particularly for industrial needs (Arnold 2001a, b).

The twenty-first century that begun with the advancement in next generation DNA sequencing technology, enzyme engineering, and systems biology provided us access to genome-level information that further facilitated the discovery of novel enzymes (Handelsman 2004). The emphasis on directed evolution that mimics natural selection in a lab condition emerged as an attractive alternative tool for developing rational strategies for engineering enzymes (Arnold 2001a, b). Moreover, the integration of synthetic biology approaches for constructing microbial cell factories for single enzyme and or multi-enzyme cascade production within a single organism emerged as a robust, cost-effective, and environmentally friendly enzymatic process. The developments in engineering of enzymes for challenging industrial conditions with high precision underscore the expanding role of microbial enzymes in industrial practices.

1.3 Pharmaceutical Applications of Microbial Enzymes

The integration of microbial enzymes into the pharmaceutical sector has undergone a remarkable evolution, significantly propelled by advancements in microbiology, biochemistry, and molecular biology. This journey can be broadly segmented into distinct periods, each characterized by the introduction of new enzyme classes and their specific applications. The late nineteenth and early twentieth centuries marked the initial foray, primarily focusing on digestive aids and rudimentary therapeutics. A pivotal moment occurred in 1894 with the commercial introduction of Takadiastase, a fungal amylase from *Aspergillus oryzae*, demonstrating the nascent potential of microbial enzymes in medicine, as a digestive aid (Samanta 2022). Concurrently, early explorations into proteases, sometimes microbially sourced, began for wound debridement or as anti-inflammatory agents. The mid-twentieth century (1940s–1970s) was heavily influenced by the antibiotic era, where the large-scale microbial fermentation for antibiotics like penicillin laid crucial groundwork for subsequent enzyme production technologies (Demain and Sanchez 2009a, b). During this period, direct therapeutic applications emerged with the development of streptokinase and streptodornase mixtures from hemolytic streptococci, used for debriding necrotic tissue and hydrolyzing DNA to improve wound healing and enhance antibiotic delivery; these were among the first injectable bacterial enzymes approved for therapeutic use (Hennigan and Lynch 2022). A significant therapeutic milestone was the increasing adoption of L-asparaginase in the 1960s, sourced from *Escherichia coli* or *Erwinia chrysanthemi*, as a direct anti-cancer agent for acute lymphoblastic leukemia (Tsegaye et al. 2024).

The late twentieth century (1980s–1990s) was revolutionized by recombinant DNA technology, enabling high-purity, high-yield enzyme production by cloning genes into microbial hosts. The first production of human insulin using recombinant *E. coli* in 1982 (https://www.gene.com/stories/cloning-insulin), though a protein therapeutic, set a vital precedent for recombinant enzymes in human medicine. Certain pathogenic microorganisms, such as the fungus Candida, use enzymes to invade host tissues and resist therapeutic interventions. As noted by Nett (2018), biofilm formation is a key virulence factor for *Candida* species. These biofilms are complex microbial communities encased in a self-produced matrix containing various proteins, DNA, and polysaccharides. Enzymes, including lipases, are secreted into this matrix to help the fungus break down host tissues, acquire nutrients, and evade the immune system. Targeting these enzymes is a potential strategy for developing new antifungal therapies to disrupt the biofilm and make the fungus vulnerable to existing drugs. Conversely, some pharmaceutical agents are designed to modulate the activity of human enzymes to treat diseases. A prime example is the class of drugs known as fibrates, which are used to treat high triglycerides and improve cholesterol levels. A Cochrane review by Räber et al. (2018) and other studies have established that fibrates work by activating peroxisome proliferator-activated receptors (PPARs), which are nuclear receptors that regulate gene expression. This activation leads to an increase in the production and activity of lipoprotein lipase (LPL). LPL is a crucial enzyme that breaks down triglycerides in the blood, effectively clearing fats from the circulation. Thus, fibrates do not directly inhibit a harmful enzyme; instead, they enhance the function of a beneficial one to improve lipid metabolism. The era also saw the increasing importance of lipases and hydrolases for stereoselective synthesis of chiral pharmaceutical intermediates, driven by the demand for enantiomerically pure drugs; examples include lipases from *Candida antarctica* (Nett 2018; Räber et al. 2018). Lipases are a diverse family of enzymes that break down triglycerides into free fatty acids and glycerol, and they are essential for fat metabolism throughout the body. Specific types of lipases are active in different tissues, each with a unique role; for example, hepatic lipase is found in the liver, while hormone-sensitive lipase is in fat cells, and lipoprotein lipase is on the vascular surface. Pancreatic lipase, in particular, is crucial for the digestion of dietary fats in the small intestine, which also allows for the absorption of fat-soluble vitamins (Kuksis 2018). Understanding lipase function is vital for comprehending the pathophysiology of diseases like fat necrosis and pancreatitis. Furthermore, lipases are clinically significant as they are targets for some cholesterol-lowering medications. Furthermore, penicillin acylase was explored for producing semi-synthetic penicillins and cephalosporins. The twenty-first century has been defined by advanced enzyme engineering techniques like directed evolution and rational design, enabling the creation of microbial enzymes with enhanced properties for complex pharmaceutical syntheses (Arnold 2001a, b). The use of transaminases for synthesizing chiral amines (Bornscheuer et al. 2012) and nitrile hydratases/nitrilases for greener production of amides and carboxylic acids (Heckmann and Paradisi 2020) has expanded. A key innovation is the rise of biocatalytic cascades, combining multiple microbial enzymes for multi-step synthesis of complex drug molecules (Chen et al. 2018). Therapeutic applications have broadened to include enzyme replacement therapies for lysosomal storage disorders and more targeted anti-cancer enzymes. Overall, microbial enzymes are central to developing more sustainable and cost-effective drug manufacturing processes, aligning with green chemistry principles (Sheldon 2016).

Enzymes are incredibly versatile biological catalysts with extensive applications across numerous fields, including organic synthesis, clinical analysis, pharmaceuticals, detergents, food production, and fermentation. Their therapeutic potential has been a significant area of exploration (Narayanan 2025), with recombinant enzymes increasingly investigated and deployed for treating various diseases (Vanderstocken et al. 2022). Enzymes of microbial origin, in particular, have found diverse uses in diagnostic assays, as biosensors, within drug delivery systems, and as direct therapeutic agents, such as antibiotics, steroids, and other pharmaceutical alternatives (Haile and Ayele 2022; Nigam 2013). For instance, asparaginase, sourced from microbes, is used to treat acute lymphoblastic leukemia, while hyaluronidase helps facilitate drug absorption. Furthermore, microbial enzymes are vital in enzyme replacement therapy (ERT) for disorders like Gaucher disease and Fabry disease (Vanderstocken et al. 2022). A historical example is Takadiastase, a fungal amylase isolated from *Aspergillus oryzae* by Jokichi Takamine in 1894, which has long been used as a digestive aid and is one of several microbial enzymes commercially produced for pharmaceutical applications (Takamine 1894; Samanta 2022).

The development of safe and effective enzyme-based human therapies has advanced significantly over the past few decades. This progress is largely attributable to our enhanced understanding of the molecular mechanisms underpinning various diseases and the identification of powerful enzymes from external sources that can offer remedial effects. Both the FDA and the EMA have approved numerous enzyme biotherapeutics, with many more currently in clinical trials. Beyond their role in enzyme replacement therapy (ERT) for genetic disorders, approved therapeutic enzymes are now broadly applied in areas such as cancer therapy, thrombolysis, and the treatment of clotting disorders, cystic fibrosis, lactose intolerance, and collagen-based conditions. The majority of these therapeutic enzymes are sourced from microbes, due to the convenience, speed, simplicity, and cost-effectiveness of microbial production and manipulation. Despite the broadened prospects for human therapy offered by microbial recombinant enzymes, several challenges persist, including high immunogenicity, protein instability, short half-life, and low substrate affinity. To address these hurdles, scientists are continuously seeking alternative enzyme sources with reduced side effects and improved activity, alongside exploring genetic modification of enzymes and developing novel delivery systems. Key strategies being investigated to mitigate toxicity, bioavailability, and cellular targeting include chemical modification, targeted and/or nanocarrier-mediated delivery, directed evolution, site-specific mutagenesis, and the creation of fusion proteins through genetic manipulation (Cioni et al. 2022).

Research into bacterial transglutaminase began in 1989 with its isolation from *Streptomyces mobaraensis*. Initially identified during a large-scale screening effort to reduce costs in food manufacturing, its exceptional activity and straightforward handling quickly established it as a robust and versatile tool for diverse biotechnological and pharmaceutical applications. While early attempts to utilize its impressive cross-linking ability primarily resulted in heterogeneous polymers, current applications involve the precise, site-specific ligation of various biomolecules, creating accurately modified hybrid constructs from two or more components. The rapidly expanding field of microbial transglutaminase-mediated bioconjugation, particularly in pharmaceutical research, is a key focus of this area. Advancements in engineering the enzyme through techniques like directed evolution and rational design are also highlighted. Furthermore, significant drawbacks of this technique, largely due to the enzyme's lack of substrate discrimination, are discussed, along with methods to overcome these limitations (Deweid et al. 2019).

Glucose oxidase (β-D-glucose:oxygen 1-oxidoreductase; EC 1.1.2.3.4) is an enzyme that catalyzes the oxidation of beta-D-glucose to gluconic acid, a process that consumes molecular oxygen as an electron acceptor and simultaneously produces hydrogen peroxide. This microbial enzyme is currently receiving considerable attention, due to its extensive applications across various sectors, including the chemical, pharmaceutical, food, beverage, clinical chemistry, and biotechnology industries. The development of novel applications, particularly in biosensors, has notably increased the demand for glucose oxidase in recent years. This review comprehensively covers the production, recovery, characterization, immobilization, and diverse applications of glucose oxidase. It details the enzyme's production via both fermentation and recombinant methods and describes various purification techniques aimed at achieving higher recovery. The review also addresses critical aspects such as enzyme kinetics, stability studies, and overall characterization. Additionally, it examines immobilized preparations of glucose oxidase and underscores the enzyme's growing impact in various industries and as an analytical tool in bioprocessing (Bankar et al. 2009).

Transglutaminases form a large family of enzymes, found both intracellularly and extracellularly, which are essential for catalyzing cross-links between protein molecules. These cross-linking properties are widely exploited in various industrial processes, especially within the food and pharmaceutical sectors, to enhance product firmness, viscosity, elasticity, and water-holding capacity. However, the significantly high costs associated with obtaining transglutaminases from animal sources have prompted scientists to actively seek new microbial sources for these enzymes (Duarte et al. 2020). Separately, the enzyme takadiastase is primarily utilized in nutraceuticals and digestive health products and has also served as a therapeutic agent. The unique activity of L-glutaminase (L-glutamine amidohydrolase EC 3.5.1.2) also presents promising potential across diverse sectors, ranging from therapeutics to the food industry. Microbial production is the preferred method for obtaining L-glutaminase over other sources, with marine microorganisms, including bacteria, yeasts, and molds, demonstrating a remarkable capacity to produce it. The broad potential of L-glutaminase in various industries has been thoroughly reviewed (Barzkar et al. 2021).

Microbial enzymes also play a crucial role in the transformation of steroids, a significant sector in the global pharmaceutical market after antibiotics, used extensively in birth control and disease treatment. Historically, steroidal active pharmaceutical ingredients (APIs) were produced through partial synthetic processes, primarily starting from sapogenins, which were famously converted into 16-dehydropregnenolone via the "Marker Degradation." Advances in traditional mutation and screening, coupled with process engineering, led to the industrial production of key steroidal intermediates such as 4-androstene-3,17-dione (AD), and rost-1,4-diene-3,17-dione (ADD), and 9α-hydroxy-androsta-4-ene-3,17-dione (9α-OH-AD). More recently, the application of genetic and metabolic engineering has resulted in highly efficient microbial strains capable of producing these precursors, allowing phytosterols to replace sapogenins as starting materials. Looking ahead, advancements in synthetic biology promise the design and construction of microbial cell factories that could industrially produce steroidal intermediates or even APIs directly from simple carbon sources like glucose. By integrating biotransformation into the synthesis of steroidal APIs, the number of reaction steps can be significantly reduced, leading to lower waste discharge and higher production efficiency, ultimately fostering a greener steroidal pharmaceutical industry (Feng et al. 2022). Microbial enzymes' ability to function under mild reaction conditions, without requiring substrate functional group protection, coupled with their extended half-life and high stereoselectivity, positions them as indispensable assets (Pandey et al. 2017). The application of microbial enzymes in the pharmaceutical industry offers several advantages, including increased efficiency, reduced waste generation, and improved product purity (Ejaz et al. 2021). These enzymes are relatively inexpensive and easy to produce on a large scale, and they also present fewer ethical issues compared to enzymes derived from animals.

1.4 Conclusion and Future Perspectives

The enzymes of microbial origin hold immense potential in the pharmaceutical sector. Further advances in genetic engineering, recombinant engineering, and synthetic biology have accelerated the developments of novel enzymes with desired biochemical characteristics such as enhanced stability, activity, and substrate specificity to perform under extreme conditions. Furthermore, the immobilization of enzymes for improving their reusability and reducing the cost as well as stability is another milestone in affordable therapeutics. However, the challenges related to immunogenicity, short half life span, regulatory issues, along with the high production costs demands efforts like AI&ML, can play valuable role in engineering efficient enzymes and developing optimized fermentation conditions. The understanding of mechanistic aspects of these enzymes will enhance the pharmaceutical potential of these enzymes for diagnostic and therapeutic applications.

References

Adrio JL, Demain AL (2014) Microbial enzymes: current status and future prospects. Appl Microbiol Biotechnol 98(2):527–535

Arnold FH (2001a) Combinatorial and computational challenges for biocatalyst design. Nature 409(6817):253–257

Arnold FH (2001b) Combinatorial chemistry and biocatalysis. Chem Eng News 79(21):38–39

Bankar SB, Bule MV, Singhal RS, Ananthanarayan L (2009) Glucose oxidase—an overview. Biotechnol Adv 27(4):489–501

Barzkar N, Sohail M, Tamadoni Jahromi S, Nahavandi R, Khodadadi M (2021) Marine microbial L-glutaminase: from pharmaceutical to food industry. Appl Microbiol Biotechnol 105(11):4453–4466

Bornscheuer UT (2002) Microbial carboxyl esterases: classification, properties and application in biocatalysis. FEMS Microbiol Rev 26(1):73–81

Bornscheuer UT, Huisman GW, Kazlauskas RJ, Lutz S, Moore JC, Robins K (2012) Engineering the third wave of biocatalysis. Nature 485(7397):185–194

Chen J, Luo Y, Van Groenigen KJ, Hungate BA, Cao J, Zhou X, Wang RW (2018) A keystone microbial enzyme for nitrogen control of soil carbon storage. Sci Adv 4(8):eaaq1689

Cherry JR, Fidantsef AL (2003) Directed evolution of industrial enzymes: an update. Curr Opin Biotechnol 14(4):438–443

Cioni P, Gabellieri E, Campanini B, Bettati S, Raboni S (2022) Use of exogenous enzymes in human therapy: approved drugs and potential applications. Curr Med Chem 29(3):411–452. https://doi.org/10.2174/0929867328666210713094722. PMID: 34259137

Crueger W, Crueger A (1990) Biotechnology: a textbook of industrial microbiology. Sinauer Associates, Sunderland

Demain AL, Sanchez S (2009a) Microbial drug discovery: 80 years of progress. J Antibiot 62(1):5–16

Demain AL, Sanchez S (2009b) Microbial production of secondary metabolites. In: Industrial biotechnology. Wiley-VCH Verlag GmbH & Co. KGaA, Weinheim, pp 1–39

Deweid L, Avrutina O, Kolmar H (2019) Microbial transglutaminase for biotechnological and biomedical engineering. Biol Chem 400(3):257–274. https://doi.org/10.1515/hsz-2018-0335. PMID: 30291779

Duarte L, Matte CR, Bizarro CV, Ayub MAZ (2020) Transglutaminases: part I—origins, sources, and biotechnological characteristics. World J Microbiol Biotechnol 36(1):15

Ejaz U, Sohail M, Ghanemi A (2021) Cellulases: from bioactivity to a variety of industrial applications. Biomimetics 6(3):44

Fasim A, More VS, More SS (2020) Large-scale production of enzymes for biotechnology uses [Review of large-scale production of enzymes for biotechnology uses]. Curr Opin Biotechnol 69:68–76. https://doi.org/10.1016/j.copbio.2020.12.002. Elsevier BV

Feng J, Wu Q, Zhu D, Ma Y (2022) Biotransformation enables innovations toward green synthesis of steroidal pharmaceuticals. ChemSusChem 15(9):e202102399

Ferrer M, Golyshin PN, Beloqui A, Nechitaylo TY (2016) The impact of microbial enzymes in the pharmaceutical industry. In: Biotechnological applications of microbial enzymes. Springer, Cham, pp 37–56

Godfrey T, West S (1996) Introduction to industrial enzymology, 2th edn. Stocholm Press, New York

Gupta R, Beg QK, Lorenz P (2002) Bacterial alkaline proteases: molecular approaches and industrial applications. Appl Microbiol Biotechnol 59(1):15–32

Haile S, Ayele A (2022) Pectinase from microorganisms and its industrial applications [Review of pectinase from microorganisms and its industrial applications]. ScientificWorldJournal 2022:1881305. https://doi.org/10.1155/2022/1881305. Hindawi Publishing Corporation

Handelsman J (2004) Metagenomics: looking at the world's microbes with a wide-angle lens. Microbe 1(2):52–58

Hassan FS, El-Fakharany EM, El-Maradny YA, Saleh AK, El-Sayed MH, Mazi W, Omer N, Abdelaziz MA, Jame R, Alatawi IS, El-Gendi H (2024) Comprehensive insight into exploring the potential of microbial enzymes in cancer therapy: progress, challenges, and opportunities: a review. Int J Biol Macromol 277(Pt 4):134535. https://doi.org/10.1016/j.ijbiomac.2024.134535. Epub 2024 Aug 5. PMID: 39111467

Heckmann CM, Paradisi F (2020) Looking back: a short history of the discovery of enzymes and how they became powerful chemical tools. ChemCatChem 12(24):6082–6102

Hennigan JN, Lynch MD (2022) The past, present, and future of enzyme-based therapies. Drug Discov Today 27(1):117–133

Kuksis A (2018) Absorption of fat-soluble vitamins. In: Fat absorption. CRC Press, Boca Raton, pp 65–86

Lorenz P, Nett M (2018) Biocatalysis in the synthesis of chiral drugs. Curr Opin Chem Biol 43:19–27

Narayanan KB (2025) Enzyme-based anti-inflammatory therapeutics for inflammatory diseases. Pharmaceutics 17(5):606

Nett JE (2018) The host's reply to Candida biofilm... Journal title not specified in search results

Nigam PS (2013) Microbial enzymes with special characteristics for biotechnological applications. Biomolecules 3(3):597–611

Pandey K, Singh B, Pandey AK, Badruddin IJ, Pandey S, Mishra VK, Jain PA (2017) Application of microbial enzymes in industrial waste water treatment. Int J Curr Microbiol App Sci 6(8):1243–1254

Räber L, Klingenberg R, Haller ML (2018) Fibrates for the prevention of secondary diseases... Cochrane Database Syst Rev

Rani R, Singh R (2017) Microbial enzymes as therapeutic agents. J Bioprocess Biorefin 1(1):1–8

Samanta S (2022) Structural and catalytical features of different amylases and their potential applications. Jordan J Biol Sci 15(2):311–337

Sheldon RA (2016) Green chemistry, catalysis and valorization of waste biomass. J Mol Catal A Chem 422:3–12

Sheldon RA, Woodley JM (2018) Role of biocatalysis in sustainable chemical production. Chem Rev 118(2):801–838

Takamine J (1894) Process of making diastatic enzyme. U.S. Patent No. 525,823. U.S. Patent and Trademark Office

Tsegaye K, Tsehai BA, Getie B (2024) Desirable L-asparaginases for treating cancer and current research trends. Front Microbiol 15:1269282

Vanderstocken G, Woolf NL, Trigiante G, Jackson J, McGoldrick R (2022) Harnessing the potential of enzymes as inhaled therapeutics in respiratory tract diseases: a review of the literature. Biomedicine 10(6):1440

Vellard M (2003) The enzyme as drug: application of enzymes as pharmaceuticals. Curr Opin Biotechnol 14(4):444–450

Yadav RP, Yadav KDS (2018) Therapeutic applications of microbial enzymes. 3 Biotech 8(1):1–10

Risk Assessment and International Biosafety Guidelines for the Enzyme Replacement Therapy in the Pharmaceutical Sector

2

Richa Salwan and Vivek Sharma

Abstract

Enzymes are indispensable tools in the pharmaceutical sector as they are being explored as a part of biological medicines. These therapeutics proteins are available in a range of size and structural complexities, spanning from relatively simple proteins like insulin or growth hormone to more intricate ones such as coagulation factors or monoclonal antibodies. Enzyme replacement therapies have been recommended by FDA to address the various diseases. The utilization of enzymes in the pharmaceutical industry is projected to undergo sustained expansion, driven by the escalating demand for sustainable and eco-friendly manufacturing processes. Nevertheless, the pharmaceutical sector must confront several pivotal challenges to ensure the accountable and successful long-term incorporation of enzymatic technologies. As the demand for the utilization of enzymes in the pharmaceutical industry continue to increase, it is crucial to thoroughly assess and manage the inherent risks associated with their exploration and application. Robust biosafety guidelines must be implemented to address the potential risks and ensure the safety and efficacy of these biocatalysts. This chapter describes the regulatory framework on the usage of biological medicines such as proteins in pharmaceutical industries with a particular emphasis on risk assessment procedures and biosafety guidelines.

R. Salwan
College of Horticulture and Forestry, Dr YS Parmar, University of Horticulture and Forestry, Neri, Hamirpur, Himachal Pradesh, India

V. Sharma (✉)
University Centre for Research and Development, Chandigarh University, Mohali, Punjab, India

© The Author(s), under exclusive license to Springer Nature Singapore Pte Ltd. 2026

V. Sharma, R. Salwan (eds.), *Pharmaceutically Relevant Microbial Enzymes*,
https://doi.org/10.1007/978-981-95-4941-2_2

Keywords

Proteins · Disorders · Treatment · Regulation

2.1 Introduction

With the advent of deoxyribonucleic acid (DNA) as genetic material, researchers discovered that specific regions of DNA contain the information necessary to direct the synthesis of particular proteins. The final product of a gene, the protein has a distinct identity and function with many serving as biocatalyst known as enzyme, while others act as structural components of cells. In particular, the proteins as enzyme are responsible for carrying out the majority of essential processes within a cell (Comfort 2000). These biocatalysts are highly specific, can operate under mild conditions, and are generally eco-friendly, making them a preferred choice for pharmaceutical applications (Meyer et al. 2012) as they offer significant advantages compared to traditional chemical synthesis methods. Enzymes for therapeutic purposes can be obtained from human, animal, plant, or microbial sources. The large-scale production of the desired protein regardless the origin, can be achieved using recombinant engineering gene either within a Prokaryotic host such as *Escherichia coli* or Eukaryotic host system such as *Pichia pastoris*. As the pharmaceutical industry continues to grow, the need for a diverse array of commercially available enzymes has become critical for enzyme-based therapies (Jaramillo et al. 2011; Meghwanshi et al. 2020).

Protein-based therapies have been proven successful in managing conditions that otherwise lack effective treatment options, including immune disorders, metabolic disorders, and certain cancers. The active enzyme or a blend of enzymes-based preparation originating from diverse sources including plants, animal tissues, or microorganisms has been formulated. The enzymes are formulated with ingredients such as diluents, preservatives, stabilizers, or other suitable substances to retain their activity. The potential of enzyme-based therapies remains largely untapped in drug discovery. The purpose of enzyme replacement is to replace the defective endogenous enzymes. However, the therapeutic utility of enzymes extends beyond replacement therapy. Other applications include uses in cancer, wound healing, microbial infections, pancreatic disorders, bacterial or viral infections, inflammatory diseases, and respiratory diseases with the option to administer the enzymes through various routes such as oral, subcutaneous, intravenous, intrathecal, intracerebroventricular, or inhalation delivery (Vanderstocken et al. 2022).

Therapeutics enzymes can be produced cost-effectively at a large scale from microbial sources, while also posing fewer ethical concerns compared to enzymes derived from animal sources. The phage enzymes with promising antimicrobial properties, such as lysins, autolysins, lysozymes, bacteriocins, endolysins, and depolymerases, have gained the attention of researchers. For example, the enzybiotics including the endolysin CF-301 are currently in clinical development for treating bloodstream infection and endocarditis caused by *Staphylococcus aureus*

(Vanderstocken et al. 2022). Bacteriophage-derived lysins, such as CF-301, consist of cell-wall-hydrolytic enzymes with 32- to 100-fold or greater demonstrated potency. Furthermore, CF-301 exhibits a synergistic effect when combined with two key human blood factors: serum lysozyme and serum albumin, which normally possess negligible antistaphylococcal activity against a prototypical methicillin-resistant *Staphylococcus aureus* strain (Indiani et al. 2019). Another enzyme known as "Adagen" used in pharmaceutical industries consists of multiple strands of monomethoxypolyethylene glycol with molecular weight of 5000 covalently linked to the enzyme adenosine deaminase of bovine intestine. This enzyme in mammals catalyzes the conversion of the purine base adenosine into inosine and ammonia, by removing an amino group from the adenosine molecule. Adagen is prescribed for enzyme therapy in individuals with severe combined immunodeficiency disorder attributable to adenosine deaminase deficiency. For the enzyme preparation is recommended for those patients who are either not eligible or where bone marrow transplantation is not recommended. The treatment is also approved for newborns or children of any age. However, the enzyme was discontinued due to its limited availability (https://reference.medscape.com/drug/adagen-pegademase-342862). The enzyme adenosine deaminase breaks down the excess adenosine present in the circulation of SCID patients, reducing the toxicity to their immune system caused by elevated adenosine levels. The success of this treatment relies on the modification of ADA with polyethylene glycol (Vellard 2003). With a recommended target dose of 15–35 μmol/h/ml, treatment reduces erythrocyte deoxyadenosine triphosphate content by approximately ≤0.005–0.015 μmol/ml in red blood cells or 1% in total adenine nucleotide content in red blood cells. In some cases, the dosage adjustments are also needed if there is evidence of antibody formation, tolerance, or changes in adenosine deaminase activity. Caution is advised when administering to patients with thrombocytopenia (https://www.vinmec.com/eng/article/uses-of-adagen-en). Similar to this, the enzyme alglucerase also known as ceredase1 has been recommended for the treatment of Gaucher disease. In individual with lysosomal storage disorders, the cell organelles known as "Lysosome" responsible for waste disposal and recycling centers cannot recycle cellular material and hence develop a range of symptoms. The French physician Philippe Gaucher's medical dissertation provided the first clinical description of the Gaucher disease as an autosomal recessive lysosomal storage disorder (https://patents.google.com/patent/US20110027254A1/en). The glucocerebrosidase-deficient cells fail to metabolize the fatty molecules present in white and red blood cell membranes. The disease predominantly targets the macrophages, a white blood cells types where the accumulation of glucocerebroside in the liver and spleen results in enlarged and impaired normal function. The treatment involves enzyme replacement therapy (ERT), through the administration of glucocerebrosidase enzyme known as "Ceredase" (Barton et al. 1991). Using genetic engineering, a modified placental glucocerebrosidase "Ceredase1" was approved in 1994. The modified enzyme obtained either from human placenta or produced in genetically engineered Chinese hamster ovary cells transfected with a GCB expression construct, referred to alglucerase or imiglucerase, respectively, have been developed (Nair and Devi 2022).

Another lysosomal disorder known as Fabry's disease is a rare genetic disorder characterized by the deficiency of α-galactosidase A which impacts the cardiovascular system. The deficiency of the enzyme led to complications such as renal dysfunction, debilitating pain, and corneal opacity. The disease is sex linked and can be easily assessed using the level of α-galactosidase A activity in leukocytes. However, the disease diagnosis is challenging in female patients, as they often have normal or near-normal residual enzyme activity (Chan and Adam 2018; Michaud et al. 2021). The treatment trail therapy of Fabry's disease have completed phase III clinical trials and its regulatory approval, as well as potential orphan drug designation for ERT against Fabry's disease is in process. The recombinant α-galactosidase expressed in Chinese hamster ovary cells or in human cells or in plant cells has been approved as orphan drug designation by the FDA in early 2003 (Eng et al. 2001; Schiffmann et al. 2001; Kakkis 2002).

ERT for various lysosomal storage (LSD) diseases is currently in experimental stage. LSDs collectively represent around 70 disorders due to mutations in proteins required for normal functioning of lysosome. Several LSDs are curable using ERT (He et al. 2012). For example, aldurazyme, a glycoprotein of ~82 kDa in size, has been approved for treating mucopolysaccharidosis Type I (MPS I) (Hampe et al. 2021). Another LSD is characterized by a deficit in α-L-iduronidase; a carbohydrate processing enzyme is responsible for metabolizing dermatan sulfate and heparan sulfate The deficiency of iduronate 2-sulfatase results in a genetic childhood disease known as Hunter syndrome due to mutations in the *IDS* gene encoding iduronate-2-sulfatase (Kakkis et al. 2001; Kakkis 2002; Stütz and Wrodnigg 2016; Demydchuk et al. 2017). Another enzyme Aryplase (BM-102) representing recombinant N-acetylgalactosamine 4-sulfatase has been approved for the treatment of mucopolysaccharidosis type VI (also known as Maroteaux-Lamy syndrome (Yogalingam 2004). Pompe's disease (as acid-maltase disease and glycogen storage disease II) is characterized by a deficiency in the enzyme α-glucosidase. This enzyme is responsible for breaking glycogen. The genetic disorder in α-glucosidase causes weakness to the heart and skeletal muscles (https://www.ninds.nih.gov/health-information/disorders/pompe-disease). The experimental study reports revealed based on recombinant enzymes appear promising. However, high doses are required for successful treatment of this debilitating disorder, and immune responses may interfere with the efficacy of the therapy in some patients (Van den Hout et al. 2001).

With a growing demand for ERT in pharmaceutical industries, the exploration of enzymes carries inherent risks must be carefully assessed and managed through the implementation of robust biosafety guidelines. Therefore, it is critical to address the potential risks associated with their use, and hence, robust biosafety guidelines are needed to ensure the safety and efficacy of these biocatalysts. This chapter attempts to provide an overview of the current international regulatory framework governing the application of enzymes in the pharmaceutical sector, with a particular focus on risk assessment procedures and biosafety guidelines.

2.2 Enzymes Replacement Therapeutics (ERT)

The pharmaceutical sector has been revolutionized by the development of novel therapeutics, including enzyme-based treatments, gene therapy, and stem cell therapies. Particularly, the ability to catalyze reactions with high selectivity and efficiency led to the widespread utilization of enzymes in the industry for various applications. Here, our discussion is restricted solely on the use of enzymes in therapeutics, excluding the mode of administration such as nebulizers, inhalers, and other delivery methods. For example, the ERT dornase alfa, which is a synthetic mucolytic deoxyribonuclease approved by the FDA in 1993 for treating cystic fibrosis. This enzyme facilitates the mucus clearance in infected lungs by targeting the breakdown of DNA. In 2020, another modified human DNase I, known as Alidornase alfa, was approved by the FDA for pulmonary sarcoidosis. Clinical trials for Alidornase alfa in individuals with cystic fibrosis showed only mild adverse symptoms, indicating that the enzyme preparation has been effective in clearing airway mucus obstruction (Vanderstocken et al. 2022). Bedside inhalation, ERTs have been developed for treating gastrointestinal diseases through oral administration (Cao et al. 1990; Basso and Şerban 2019; Meghwanshi et al. 2020; Štreimikytė et al. 2022; Liang et al. 2022). Beyond respiratory conditions, ERT have been developed for the treatment of gastrointestinal diseases through oral administration as well as for cardiovascular diseases. ERT have also been utilized for the successful treatment of cardiovascular conditions. For instance, the thrombolytic agent streptokinase and urokinase have been employed to manage acute myocardial infarction and thrombotic disorders. Additionally, the immobilization of enzymes has been used as biosensors for the detection of clinically important analytes including glucose, lactate, and urea.

2.3 Biosafety Guidelines for the Usage of Enzymes

The utilization of enzymes in therapeutics in the pharmaceutical industry require the implementation of robust biosafety protocols to mitigate potential risks. These biosafety protocols constitute a framework of guiding principles, practical procedures, and established practices intended to mitigate the risk of inadvertent release or unintended exposure to biological agents and toxins that may pose threats to human, animal, or plant wellbeing (Richmond and McKinney 1999; Beeckman and Rüdelsheim 2020). These biosafety guidelines outline the protocols and practices necessary for the safe management, containment, and disposal of potentially hazardous biological materials including enzymes, in order to safeguard laboratory personnel, the environment, and the general public (Burnett et al. 2009). Biosafety guidelines constitute a comprehensive framework of policies, protocols, and procedures that must be strictly followed by personnel working in various facilities handling microbiological agents, such as bacteria, viruses, parasites, fungi, prions, and other related microbiological products. These guidelines are mandatory for clinical and microbiological laboratories, biomedical research centers, teaching and training laboratories, and other healthcare institutions. The primary purpose of these guidelines is to ensure the proper management and regulation of biosafety programs and practices implemented across the organization (Bayot and Limaiem 2019). The core components of biosafety guidelines typically encompass a multifaceted approach customized to the specific facility. This may include comprehensive bio-risk evaluation and identification, as well as detailed protocols governing operational procedures, physical infrastructure design, equipment administration, medical surveillance, personnel training, and safety measures for hazardous materials, fire, radiation, and electrical systems. Depending on the organization, supplementary elements such as facility commissioning and certification guidelines may also be incorporated (Bayot and Limaiem 2019).

The pharmaceutical industries are required to implement rigorous containment strategies to mitigate the risks associated with the handling and processing of enzymes. This may encompass the utilization of specialized equipment, such as biosafety cabinets, and the establishment of stringent protocols for the containment of any enzyme spills or accidental releases. Furthermore, appropriate decontamination and waste management procedures are crucial to preclude the environmental discharge of enzymes and safeguard the wellbeing of workers and the surrounding community (Fig. 2.1).

Continuous monitoring and surveillance of enzyme-based pharmaceutical products is essential for identifying and addressing any emerging risks or safety concerns. Robust monitoring programs should be implemented to track the performance, stability, and environmental impact of enzymes throughout their lifecycle. Periodic safety evaluations and risk assessments can assist the pharmaceutical industry in anticipating and mitigating potential issues, thereby ensuring the ongoing safe utilization of enzymatic technologies. The pharmaceutical industry must develop

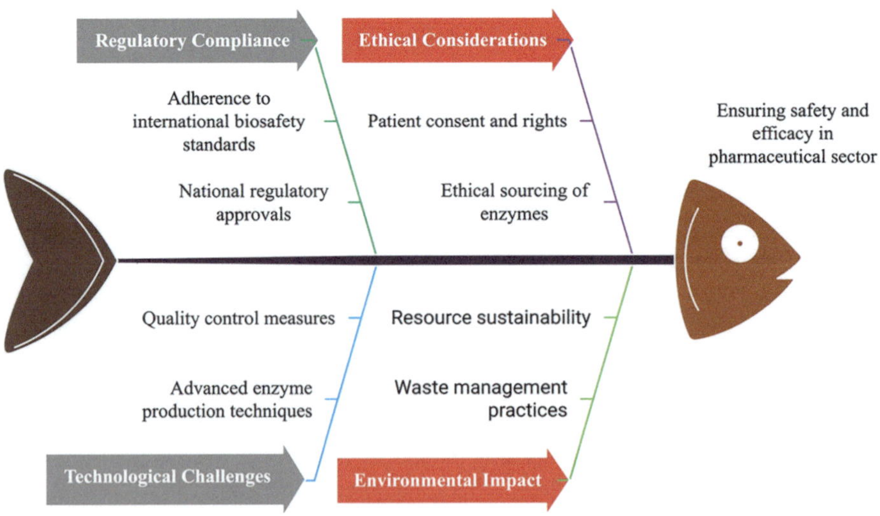

Fig. 2.1 Risk assessments and biosafety guidelines overview in enzyme replacement therapy

comprehensive emergency plans to address any incidents or accidents involving enzymes. These plans should outline clear protocols for the containment and decontamination of affected areas as well as procedures for the safe transportation and disposal of contaminated materials. Regular training for emergency response personnel is needed to ensure that the industry is prepared to respond effectively to any enzyme-related incidents (Lu et al. 2009; Regil and Sandoval 2013; Švarc et al. 2019; Meghwanshi et al. 2020).

The applications of enzymes in the pharmaceutical sector demands safety, sustainability, and ethical considerations. Ensuring the competency of workers is essential for the safe and responsible exploration of enzymes in the pharmaceutical sector. Comprehensive training programs should be implemented to educate workers on the proper handling, storage, and disposal of enzymes, as well as the appropriate safety and containment protocols. Regular auditing and compliance monitoring are crucial to ensure that the pharmaceutical industry is adhering to all relevant international biosafety guidelines and regulations (Tao and Xu 2009; Nair et al. 2014). Third-party audits and internal compliance reviews can help identify areas for improvement and ensure the ongoing effectiveness of the industry's risk assessment and management strategies. Ethical guidelines and stringent mechanisms such as independent ethical committees and regulatory bodies need to ensure that the usage of enzyme aligns with ethical principles and societal values (Fig. 2.1).

Maintaining the highest standards of research integrity are of paramount importance for the exploration of enzymes in the pharmaceutical industries. For that, the collaborative partnerships, engagement with stakeholders including regulatory bodies, environmental organizations, and local communities and open dialogues can empower the industry to ultimately ensuring the responsible and sustainable development of enzyme-based pharmaceutical products. The adherence to standard protocols for handling maintenance of transparent and accurate record during evaluation of enzymes is needed. The integrity and reliability of research findings must be established through robust quality assurance and quality control. For that, the pharmaceutical industry must align with international biosafety guidelines, implement comprehensive risk assessment and management strategies, and prioritize environmental and occupational health considerations. The alignment with international biosafety standards will bolsters the safe and responsible development of enzymes in accordance with the industrial demands (Fig. 2.1).

At the organization level, the research group working on enzymes and other biological systems must adhere to guidelines provided by the International Organization for Standardization (ISO 35001:2019). The standard provides recommendation for risk assessment, containment, and other preventive measures associated upon exposure. The compliance with international and national regulatory frameworks ensures the highest levels of biosafety in the enzyme exploration for pharmaceutical industries. For pharmaceutical enzymes, the guidelines must be in adherence to strict regulatory guidelines established by agencies like the U.S. Food and Drug Administration (FDA) and the European Medicines Agency (EMA), responsible for the development, testing, and approval of enzyme-based pharmaceutical products. These regulatory authorities mandate rigorous risk assessments, laboratory testing, and clinical trials to ensure the safety and effectiveness of enzyme-based therapies prior to granting commercial approval. Unwavering adherence to these regulatory guidelines is imperative for the pharmaceutical industry to responsibly explore the potential applications of enzymes. Furthermore, the exploration of enzymes in the pharmaceutical industries involves a complex regulatory landscape, which varies across different countries and regions (Dec and Bollag 2000; Lu et al. 2009; Porta et al. 2010; Raghavendra et al. 2018). The pharmaceutical industry's utilization of enzymes is governed by an intricate framework of international regulations and biosafety guidelines. Pharmaceutical companies are obligated to adhere to relevant safety protocols, including those delineated by the World Health Organization and the Organisation for Economic Co-operation and Development (Regil and Sandoval 2013; Nair et al. 2014). Additionally, the international organizations, such as the World Health Organization (WHO) and the United Nations (UN), have approved detailed and comprehensive biosafety guidelines for the handling, transportation, and disposal of potentially hazardous biological materials, including enzymes. These guidelines address critical aspects like risk assessment, containment measures, and emergency response protocols, ensuring that the exploration of enzymes in the pharmaceutical industry is undertaken in a safe and responsible manner (Švarc et al. 2019; Meghwanshi et al. 2020). Adherence to these regulatory and biosafety standards is imperative for the pharmaceutical sector to mitigate the risks associated with enzymatic technologies and safeguard the well-being of workers, consumers, and the environment (Dec and Bollag 2000; Sharma et al. 2006; Porta et al. 2010). The safe and responsible management of enzyme transportation and disposal are critical components of the risk assessment and biosafety framework for the pharmaceutical industry. Appropriate methods for the manufacturing, transport of enzymes packaging, and labeling and disposal that adhere to regulatory frameworks are required to prevent any accidental damage to the environment or human health.

The environmental concerns warrant comprehensive assessment of full life cycle of enzymes starting from manufacturing to disposal. The concerns such as release and accumulation of enzymes into aquatic bodies and soils, the mechanism of their disposal, and any associated potential effects on non-target organisms must be rigorously evaluated (Lu et al. 2009; Nair et al. 2014; Švarc et al. 2019; Meghwanshi et al. 2020). The waste management, disposal strategies, and green processes for the production of enzymes that can help mitigate the environmental risks associated must be preferred. Effective risk communication and management strategies are crucial for the successful and responsible investigation of enzymes in the pharmaceutical industry. Pharmaceutical companies should clearly communicate the potential risks and benefits of enzyme-based products to key stakeholders, such as regulators, healthcare professionals, and the general public. Transparent and open communication can foster trust and facilitate the adoption of these technologies, while also addressing any concerns or misconceptions that may arise (Lu et al. 2009; Nair et al. 2014; Švarc et al. 2019; Meghwanshi et al. 2020) (Fig. 2.1).

Overall, the usage of enzymes in the pharmaceutical industries must follow sustainable and environmentally friendly practices (Nair et al. 2014). Furthermore, the pharmaceutical industry must look into the consideration of the long-term sustainability goals and focus on the source and supply as well as environment risk associated with their release (Regil and Sandoval 2013; Nair et al. 2014; Švarc et al. 2019; Meghwanshi et al. 2020).

2.4 General Considerations for the Applications of Enzymes in Human Welfare

Enzyme-based products utilized in food and pharmaceutical manufacturing contain active enzymes responsible for their intended technical effects. These preparations may consist of a single enzyme or a blend of multiple enzymes. The enzymes present can originate from various biological sources including plants, animal tissues, or microorganisms. These enzymes are formulated with intentionally added ingredients, such as diluents, preservatives, and stabilizers. Enzyme preparations may also contain residual components from the enzyme source and the manufacturing process, such as remnants of the fermentation medium used to cultivate the enzyme-producing microorganism. Enzyme preparations are typically used at low concentrations and are often removed or inactivated in the final product. For instance, enzymes employed in the conversion of starch to glucose or high fructose corn syrup are eliminated during the purification of the syrup, while enzymes used in baked goods are denatured by the high temperatures during baking. These enzyme-based products can be regulated as secondary direct food additives under Title 21 of the Code of Federal Regulations, Part 173.

Enzymes utilized in food and pharmaceutical processing are derived from animal tissues, plant materials, and microorganisms. These sources must adhere to the guidelines specified in the "Enzyme Preparations" monograph of the current Food Chemicals Codex edition. Animal tissues used to produce enzymes should meet applicable US meat inspection requirements and be handled following good hygienic practices. Plant materials and culture media employed in enzyme production should not contain components that could leave harmful residues in the final food product under normal use conditions. The fermentation process must be conducted under controlled conditions to prevent contamination by microorganisms that could be a source of toxic or undesirable substances. The microorganisms employed in enzyme production must be taxonomically identified and demonstrated to be nonpathogenic and nontoxigenic. However, certain strains typically considered nontoxigenic may have the capability to generate toxins when cultured under conditions conducive to toxin synthesis. In cases where such microorganisms serve as enzyme sources, the fermentation parameters should be adjusted to prevent toxin formation, and appropriate testing should be conducted to ensure the final enzyme preparations are free of unsafe toxin levels. Alternatively, these microorganisms may be genetically modified, through either classical mutagenesis or genetic engineering, to inactivate the biochemical pathways involved in toxin synthesis. Comprehensive information regarding the identity, safety, and any prior use in food or food ingredient production should be provided for all microorganisms utilized as enzyme sources.

Detailed information should be provided for enzymes derived from genetically modified microorganisms. These microorganisms must be thoroughly characterized with respect to any introduced DNA, including the source and identity of the gene(s) encoding the enzyme of interest, as well as any other genes and regulatory DNA sequences necessary for gene expression. The enzyme-encoding genes may originate from known organisms, unidentified environmental samples, or be generated through molecular evolution techniques such as gene shuffling. Alternatively, the genes may be synthesized or modified using traditional or site-specific mutagenesis to adapt the enzyme properties for specific food application conditions or to enhance production. The host microorganisms can also be genetically modified by inactivating or deleting certain endogenous genes, for example, to prevent the synthesis of potentially harmful secondary metabolites or to minimize the production of interfering enzymes. A comprehensive description of all the approaches and steps involved in the production of enzymes from genetically modified microorganisms should be provided.

2.5 General Considerations for the Enzyme Preparation

Successful development of enzymes for pharmaceutical use necessitates comprehensive understanding of their properties and potential risks. Extensive research has been undertaken to identify and characterize enzymes that may be appropriate for pharmaceutical applications (Dinkel et al. 2003; Lu et al. 2009; Porta et al. 2010; Meghwanshi et al. 2020). Studies have examined methods to enhance the stability and viability of enzymes for biomedical and industrial applications. Researchers have demonstrated that incorporating enzymes into silk-based matrices can significantly improve their long-term stability and activity, enhancing their potential for use in pharmaceutical products. Similarly, investigations into the application of enzymes in bioremediation have underscored the need for stabilization techniques such as immobilization or gel coating, to prevent losses of enzymatic activity under challenging environmental conditions. By understanding the physicochemical properties, stability, and potential risks of enzymes, pharmaceutical companies can make informed decisions about their suitability and develop appropriate risk mitigation strategies.

Enzyme preparations intended for pharmaceutical use should adhere to the purity requirements outlined in the "Enzyme Preparations" monograph of the current or sixth edition of the Food Chemicals Codex. Applicants should analyze multiple batches of the enzyme preparation to demonstrate compliance with these specifications. The standard analytical methods referenced in the monograph should be employed, while any non-standard methods developed by the manufacturer should be validated, described, and the activity unit defined. Enzyme preparations derived from microbial sources must be free of antibiotics, toxins, and transformable DNA encoding protein toxins or therapeutic antibiotic-inactivating proteins. If the absence of these substances cannot be established based on the production strain's characteristics, appropriate testing should be conducted to ensure the enzyme preparation does not contain these substances at biologically significant levels.

Researchers have explored approaches to enhance the stability and activity of enzymes, which is crucial for their successful application in biomedical and industrial domains (Lu et al. 2009). For example, studies have shown that incorporating enzymes into silk-based matrices can significantly improve their long-term stability and activity, thereby enhancing their viability for pharmaceutical use. Similarly, investigations into the use of enzymes in bioremediation have emphasized the need for stabilization techniques such as immobilization or gel coating to mitigate the loss of enzymatic activity under challenging environmental conditions (Dec and Bollag 2000). By thoroughly understanding the physicochemical properties, stability, and potential risks associated with enzymes, pharmaceutical companies can make informed decisions about their suitability and develop appropriate risk mitigation strategies for their utilization (Sharma et al. 2006; Lu et al. 2009).

Commercial enzyme preparations utilized in food processing commonly include the active enzyme that drives the desired chemical transformation, as well as supplementary ingredients such as stabilizers, preservatives, or diluents. These enzyme preparations may additionally contain residual components originating from the organism and manufacturing process. To comprehensively characterize the enzyme preparation, the following details should be provided:

- Details regarding the identities and concentrations of any diluents, stabilizers, preservatives, or other supplementary substances incorporated into the enzyme formulation.
- Comprehensive information on any secondary enzymes that may be present in the enzyme preparation, originating from the production organism.
- Detailed data on the residual content of other metabolites derived from the production organism, as well as any substances utilized in the enzyme isolation and purification processes.
- The total organic solids content for both the commercial enzyme preparation and the enzyme batch employed in toxicological studies.

The production of enzyme preparations from microorganisms should be thoroughly documented, detailing the fermentation process and all necessary steps and controls to ensure proper growth conditions, culture purity, and genetic stability. The isolation of the enzyme, including any chemical or physical treatments and associated quality control measures, should also be described. Additionally, all materials employed during the fermentation and subsequent downstream processing should be identified and demonstrated to be appropriate for use in food applications.

Food additive petitions or Generally Recognized as Safe (GRAS) notices for enzyme preparations should provide information on the presence of proteins derived from major food allergens, as defined by the Food Allergen Labeling and Consumer Protection Act of 2004. FALCPA requires that the labels of all FDA-regulated foods containing a major food allergen (milk, egg, peanuts, tree nuts, soybeans, wheat, fish, or crustacean shellfish) disclose the specific food source, unless the ingredient has been exempted through a petition or notification process. If an enzyme preparation contains proteins from a major food allergen and is intended for use in foods that typically do not contain that allergen, the preparation may fall under FALCPA regulations unless exempted. For example, an enzyme preparation formulated with wheat flour and intended for use in non-wheat containing foods would require disclosure or an exemption (https://www.fda.gov/regulatory-information/search-fda-guidance-documents/guidance-industry-recommendations-submission-chemical-and-technological-data-food-additive-petitions). Additional guidance on food allergen labeling requirements is available in FDA documents.

Enzymes preparation referred as secondary direct food additives under "Title 21 of the Code of Federal Regulations, Part 173 (21 CFR 173)" may be either regulated through a premarket approval process or determined to be GRAS. In both cases, the safety determination is restricted to the intended conditions of use of an enzyme preparation. GRAS verifies the claim that the enzyme preparation must be safe, generally and accepted by the scientific community. The detailed technical information such as identity, method of isolation, manufacture, specifications and use levels, dietary exposure, and other toxicological details must be mentioned. The information in GRAS notice is similar to food additive petition.

For FDA, the summary of data and information in GRAS notices is sufficient, whereas the raw data and detailed records may be submitted upon request. For FDA approval, the proprietary information of FDA discourages GRAS notices. However, if GRAS notice consists of any proprietary information, then the reason of confidentiality needs to be maintained. Thereafter, after FDA or any other agency approval is required that the information may be exempted from public disclosure under 21 CFR part 20, or not.

2.6 Importance of Risk Assessment in the Pharmaceutical Sector/Risk Assessment for Enzymes in Pharmaceutical Sector

Risk assessment is a crucial component of biosafety protocols, as it enables the identification, evaluation, and mitigation of potential risks associated with the exploration and use of enzymes in the pharmaceutical sector (Beeckman and Rüdelsheim 2020). The risk assessment process takes into account the characteristics of the enzyme, the intended use, the potential for exposure, and the potential consequences of any adverse events. The production of biological pharmaceuticals is generally more complex than chemically synthesized. Frequently, these biological medicines are manufactured using advanced biotechnology approaches, including the application of complex cellular systems and recombinant DNA technology. Accordingly, the European Union's regulatory framework imposes rigorous frameworks that govern the manufacturing of all medicinal products.

As per the Medicinal Products under Directive 2001/83/EC, biopharmaceuticals are regulated by the basic law that regulates medicinal products for human use in the

EU. A notification issued by the concerned authorities must be issued for importing medicinal products for human use, in accordance with Article 46b of Directive 2001/83/EC European Union. The confirmation is subjected to following guidelines:

1. The pharmaceutical ingredients must adhere to Good Manufacturing Practices outlined in EU directives or equivalent standards.
2. The manufacturer is under strict, regular, and transparent controls with effective enforcement of Good Manufacturing Practice, including repeated and unannounced inspections, to ensure public health protection equivalent as per EU.
3. The manufacturer needs to adhere to the Guidance document for Issue of Written Confirmation issued by the Central Drugs Standard Control Organization.
4. The authorized exporter shall produce a written confirmation as required by the Drug Regulatory Authority. This written confirmation, unless sooner suspended or cancelled, shall be valid for a period of 3 years.
5. The written confirmation will be withdrawn in the event of non-compliance with the stated standards.
6. Any non-compliance observed during inspections conducted by Local or International Drug Authorities shall be forwarded to the issuing office within 7 days of receipt of the report.
7. Any drug found to be of non-standard quality shall be reported to the issuing office within 7 days of receipt of the report.

I. *Importance of International Harmonization*

   1. The use of fixed-dose combinations of laxatives and/or antispasmodic drugs in enzyme-based pharmaceutical preparations.
   2. International harmonization is crucial for ensuring the effective regulation of pharmaceuticals, including pharmaceutical enzymes.

As per Indian, "Pharmacopoeia" which is the official compendium of standards for, established under the Drugs and Cosmetics Act of 1940 and associated regulations from 1945, the active ingredients other than enzymes, antibiotics and vitamins, in medicines shall be not be below 90% and over 110% of the labelled content. For enzymes and vitamins, only the lower limit of 90% shall be applicable (https://naco.gov.in/sites/default/files/Drug%20%26%20Cosmetic%20Act%201940_0.pdf; https://ipc.gov.in/images/GD-01-Legal_Status_of_IP.pdf) (The Drugs and Cosmetics Act 1940; The Drugs and Cosmetics Rules 1945). The Codex Committee on Food Additives compiled a priority list of 84 substances, 29 of which were enzyme preparations produced by microorganisms, that were proposed for evaluation by the Joint FAO/WHO Expert Committee on Food Additives. To aid in addressing these requests, the 86th Working Group examined the existing safety evaluation process for enzyme preparations (https://www.fao.org/food/food-safety-quality/scientific-advice/jecfa/jecfa-additives/en/).

Pharmaceutical companies that market biosimilar products are obligated to closely monitor and report any suspected adverse drug reactions. Regulatory authorities review these reports to detect potential undesirable effects, which are then rigorously evaluated by scientific committees to determine if any corrective measures are warranted. Moreover, all new biological medicines approved in the European Union after January 1, 2011, are subject to "enhanced monitoring" and are included in a special surveillance list. These products are closely scrutinized by regulatory bodies, particularly during the initial years following authorization, due to the novelty of the active substance or limited long-term data.

Monitoring the long-term or delayed adverse events associated with biological medicines is crucial, as the principles align with those of small-molecule pharmaceuticals. However, identifying and characterizing the long-term adverse drug reactions of biological medicines may pose challenges when relying solely on spontaneous reporting. Consequently, supplementary pharmacovigilance efforts may be necessary in specific cases to enhance the monitoring process. The traceability of biological medicines is pivotal for effective safety monitoring. European Union regulations mandate that all medicines have a distinctive invented name and active substance name. For biological medicines, identification and tracing necessitate the use of the tradename and batch number, particularly when multiple products with the same active ingredient are available. This ensures that any product-specific safety concerns can be properly attributed. Healthcare professionals play a crucial role in understanding a medicine's safety profile during clinical use. Biological medicines are approved based on an acceptable safety profile and should be used in accordance with the summary of product characteristics and package leaflet recommendations. If a suspected adverse drug reaction is identified for a biological medicine, healthcare professionals must report it, including the tradename and batch number. They should also report any suspected adverse reactions for biosimilars, even if the reaction is already listed for the reference medicine. The tradename, international nonproprietary name, and batch number for a biological medicine can be found on the product packaging. The summary of product characteristics now includes a statement reminding healthcare professionals to diligently record the tradename and batch number in patient records.

The European Union's Summary of Product Characteristics serves as a crucial resource for healthcare professionals, providing essential information and guidance for the prescription and patient counseling of medicinal products. As outlined in Sect. 5.1, the SmPC indicates whether a medicine is a biosimilar, with detailed supporting information available on the European Medicines Agency's website. Within the EU regulatory framework, the SmPC of a biosimilar is closely aligned with that of the reference medicinal product. While the biosimilar's SmPC specifies the name of the active substance rather than the reference product's tradename, details of the biosimilar's clinical studies and the reference product's tradename can be found in the EMA's publicly accessible assessment report. Notably, a biosimilar may be approved for a subset or the entirety of the reference product's authorized indications, as the marketing authorization holder may selectively pursue certain indications. Healthcare professionals should verify that the biosimilar is authorized for the intended clinical application. In instances where the biosimilar does not cover all the reference product's indications, the SmPC will not include efficacy data for those additional uses, although safety information will still be reflected. The European Medicines Agency publishes the European Public Assessment Report, a comprehensive set of documents, for each medicinal product approved through the centralized procedure, including biosimilars.

2.7 Enzyme Preparations Containing Allergenic Ingredients

The assessment and management of risks associated with the exploration and utilization of enzymes in the pharmaceutical sector are critical. Enzymes, being proteinaceous in nature, hold the potential to elicit allergic reactions in individuals who have been extensively exposed to them. Additionally, the employment of genetically modified organisms for enzyme production raises concerns regarding biosafety, as the possibility of unintended consequences and environmental release must be thoroughly considered. To effectively mitigate these risks, comprehensive risk assessment protocols are essential. This entails a thorough evaluation of the enzyme's physicochemical properties, its propensity to trigger immune responses, and the safety of the production process and host organism. Moreover, employees handling enzymes must be appropriately trained in safety protocols to minimize exposure and potential sensitization. Pharmaceutical manufacturers in the European Union are mandated to possess a valid manufacturing license and strictly adhere to Good Manufacturing Practice standards to ensure the quality of their medicinal products.

Regulatory agencies within the European Union routinely inspect pharmaceutical manufacturing facilities to ensure compliance with Good Manufacturing Practice standards. This requirement extends to non-EU manufacturers, importers, and wholesalers involved in any stages of the production process that occur outside the EU, who must also adhere to these rigorous guidelines and undergo regular inspections. Moreover, the GMP regulations for biological medicinal products have been adapted to address their unique characteristics. The human immune system has the ability to recognize and respond to foreign proteins. Biopharmaceuticals typically elicit minimal or negligible immune responses, and any adverse immune-related reactions are generally mild. However, in rare instances, an immune response to a biological drug could be severe and potentially life-threatening. Since the first biosimilar was introduced for clinical use in 2006, a growing number of biosimilar products have been authorized and safely utilized within the European Union. Excluding immunological reactions, the majority of adverse drug events can be anticipated based on the pharmacological activity and are observed with both the reference medicine and the biosimilar. To date, none of the over 50 biosimilars approved in the EU have been withdrawn or suspended due to safety or efficacy concerns.

When evaluating the potential allergenicity of an enzyme, relevant factors to consider include the source of the enzyme, a comparison of its amino acid sequence to known allergens, and its susceptibility to digestion by the enzyme pepsin. These characteristics, aligned with current scientific understanding, can provide insight into the likelihood of the enzyme eliciting an allergic response. Additionally, any other pertinent information should be taken into account during the assessment process.

2.8 Hazard Identification in Enzyme Exploration

Prior to commercial applications enzymes for pharmaceutical industries, the potential risk related to human health and the surrounding ecosystem associated with their usage must be evaluated and properly managed. The disposal of enzymes into the environment at production or at any other stage may cause detrimental effects to the environment. These enzymes can also behave as an allergen and hence may lead to immunogenic response. Furthermore, in some cases, the inherent toxicity may lead to adverse reactions in individuals exposed to these enzyme preparations. Therefore, the environmental impact must be assessed prior to their exploration for pharmaceutical applications (Sharma et al. 2006). The fate of enzyme preparations in the environment such as ability to persist or degrade must be evaluated for the potential ecological impacts. Also, the assessment of the eco-toxicological impacts of the enzyme preparations against non-targeted organisms present in water bodies or terrestrial habitats is necessary to avoid the any potential adverse risk. The environmental risks can be minimized using rigorous assessments.

The Food, Drug, and Cosmetic Act exempts substances that are generally recognized as safe from the definition of a food additive. A substance can be considered GRAS if there is evidence of its safety and a consensus among qualified experts. On the other side, the assessments of European Union for food enzymes declares that the "food enzyme" and "food enzyme preparation" needs a comprehensive safety evaluation prior to the authorization in the industry. The safety criteria evaluate the safety of the enzyme preparations, the production source, the process of manufacturing, and residual assessments (Silano et al. 2020; Jensen and Eigtved 1990). Some commercially available enzyme preparations may contain allergenic ingredients, such as microbial proteins or residual components from the fermentation medium used to cultivate the microorganism producing the enzyme. These allergenic substances can pose a risk to individuals with preexisting sensitivities, potentially causing adverse reactions upon exposure. To address this risk, manufacturers must carefully assess the composition of enzyme preparations and implement rigorous allergen management systems. This includes identifying potential allergens, monitoring exposure levels, and providing clear labeling and warnings for products containing allergenic materials.

Recognizing the importance of responsible enzyme development, several international organizations have established biosafety guidelines to govern enzyme research and applications in the pharmaceutical industry. The World Health Organization has published guidelines on the safe handling and containment of biological agents, which serve as a foundation for national regulatory frameworks. Additionally, the Organization for Economic Cooperation and Development has developed a set of principles for the safety assessment of microbially derived enzymes, focusing on potential toxicity, allergenicity, and environmental impact

(Nair et al. 2014). These guidelines emphasize the need for comprehensive testing, risk mitigation strategies, and oversight to ensure the safe and sustainable use of enzymes in pharmaceutical applications (Grundtvig et al. 2018; Nair et al. 2014). Adherence to these international standards is crucial for companies and researchers seeking to explore the potential of enzymes in the pharmaceutical sector. By prioritizing risk assessment and following established biosafety protocols, the industry can harness the benefits of enzymatic technologies while ensuring the safety of workers, consumers, and the environment (Dec and Bollag 2000; Porta et al. 2010; Švarc et al. 2019; Meghwanshi et al. 2020).

2.9 Case Study

The U.S. Food and Drug Administration has approved Lamzede, the first enzyme replacement therapy for the treatment of non-central nervous system manifestations associated with the rare genetic disorder alpha-mannosidosis. Alpha-mannosidosis is characterized by a deficiency or dysfunction of the lysosomal alpha-mannosidase enzyme due to mutations in the MAN2B1 gene. This genetic condition can lead to a range of symptoms, including mild to moderate intellectual disability, hearing loss, weakened immune system, distinctive facial features, skeletal abnormalities, and muscle weakness. Lamzede is designed to act similarly to the alpha-mannosidase enzyme, thus restoring normal cellular activity in patients. The recommended dosage for Lamzede is a 10 mg intravenous injection administered once per week.

Alpha-mannosidosis is a rare genetic disorder characterized by a deficiency or dysfunction of the lysosomal alpha-mannosidase enzyme due to mutations in the MAN2B1 gene. This can lead to a range of symptoms, including mild to moderate intellectual disability, hearing loss, weakened immune system, distinctive facial features, skeletal abnormalities, and muscle weakness. The condition affects approximately 1 in every 500,000 people worldwide. The efficacy of the enzyme replacement therapy Lamzede was evaluated in a phase 3 clinical trial involving adult and pediatric patients with alpha-mannosidosis. The study was a multicenter, randomized, double-blind, placebo-controlled trial that assessed Lamzede's efficacy over 52 weeks at a dose of 1 mg/kg administered weekly as an intravenous infusion. A total of 25 patients were enrolled, including 13 adults and 12 children, all of whom were White. Fifteen patients received Lamzede, while 10 received a placebo. The efficacy results, measured by clinical endpoints such as the 3-min stair climbing test, 6-min walking test, and forced vital capacity, all favored the Lamzede group and were supported by a reduction in serum oligosaccharide concentration. The most common adverse reactions to Lamzede are hypersensitivity reactions, including anaphylaxis, a severe and potentially life-threatening allergic reaction. Appropriate medical support measures should be available during Lamzede administration, and if a severe hypersensitivity reaction occurs, the treatment should be discontinued immediately, and appropriate medical intervention should be initiated. In cases of severe hypersensitivity, healthcare professionals may consider a desensitization procedure to Lamzede. The development of the document involved a collaborative effort between the European Medicines Agency, the European Commission, and scientific experts from European Union member states. Based on feedback and input from healthcare professional organizations, the EU has been actively involved in the regulation of biosimilars, having approved the highest number worldwide over the past 10 years, gaining considerable experience in their use and safety.

Improving the stability and catalytic activity of enzymes under industrial conditions, developing efficient processes along with refinements in the purification and mitigating potential risks associated with enzyme release and disposal is required for enzyme manufacturing and purification. The pharmaceutical industry's exploration of enzymes must be underpinned by a commitment to ongoing improvement and innovation. Regular risk assessment, compliance monitoring, and stakeholder engagement will be crucial for the industry to stay ahead of emerging challenges and maintain the trust of the public and regulatory authorities. In conclusion, the responsible exploration of enzymes in the pharmaceutical sector requires a multifaceted approach that balances innovation, safety, sustainability, and ethical considerations. By upholding robust standards and fostering collaborative partnerships, the pharmaceutical industry can harness the full potential of enzymatic technologies while ensuring the responsible development of enzyme-based pharmaceutical products. Prospective therapeutic enzyme superoxide dismutase, the most important detoxifying enzyme present in cells, transforms the highly toxic superoxide anion to moderately toxic hydrogen peroxide. The same is true for catalase, another antioxidant that converts hydrogen peroxide to water and oxygen. Surprisingly, these enzymes have been shown to prolong the life of (the nematode) *Caenorhabditis elegans* and this effect may translate to mammals. Future versions of these enzymes may help to reduce organ injury in hemorrhagic shock and may have utility in tissue repair and wound healing (Adrio and Demain 2014).

2.10 Conclusion and Challenges

The utilization of enzymes in the pharmaceutical realm presents both prospects and obstacles. Robust regulatory structures and adherence to international biosafety guidelines are indispensable for the accountable advancement and employment of enzymatic technologies. Comprehensive risk assessments, encompassing the identification and characterization of enzymes, as well as meticulous environmental impact evaluations, are pivotal steps in ensuring the safe and sustainable exploration of enzymes within the pharmaceutical industry. By proactively addressing potential hazards and implementing appropriate risk mitigation strategies, the pharmaceutical sector can harness the benefits of enzymatic technologies while upholding the highest standards of safety and environmental stewardship. Adherence to these guidelines is essential for safeguarding worker health, securing public safety, and maintaining the overall integrity of the pharmaceutical industry's enzyme-based research and production activities.

References

Adrio J, Demain AL (2014) Microbial enzymes: tools for biotechnological processes. Biomolecules 4(1):117–139. https://doi.org/10.3390/biom4010117

Barton N, Brady R, Dambrosia J, Bisceglie AD (1991) Replacement therapy for inherited enzyme deficiency—macrophage-targeted glucocerebrosidase for Gaucher's disease. N Engl J Med 324(21):265–273. https://doi.org/10.1142/9789814273596_0013

Basso A, Şerban S (2019) Industrial applications of immobilized enzymes—a review. Mol Catal 479:110607. https://doi.org/10.1016/j.mcat.2019.110607

Bayot ML, Limaiem F (2019) Biosafety guidelines

Beeckman DSA, Rüdelsheim P (2020) Biosafety and biosecurity in containment: a regulatory overview. Front Bioeng Biotechnol 8:650. https://doi.org/10.3389/fbioe.2020.00650

Burnett LC, Lunn G, Coico R (2009) Biosafety: guidelines for working with pathogenic and infectious microorganisms. Curr Protoc Microbiol 13(1):1A-1. https://doi.org/10.1002/9780471729259.mc01a01s13

Cao S, Zhao Q, Ding Z, Ma L, Tao Y, Wang J, Feng Y, Cheng Y (1990) Chemical modification of enzyme molecules to improve their characteristics. Ann N Y Acad Sci 613(1):460–467. https://doi.org/10.1111/j.1749-6632.1990.tb18201.x

Chan B, Adam DN (2018) A review of Fabry disease. Skin Therapy Lett 23(2):4–6. PMID: 29562089.

Comfort N (2000) Book Review: A history of molecular biology Michel Morange, Matthew Cobb. Q Rev Biol 75(3):303–303. https://doi.org/10.1086/393506

Dec J, Bollag J (2000) Use of enzymes in bioremediation. American Chemical Society, pp 182–193. https://doi.org/10.1021/bk-2001-0777.ch010

Demydchuk M, Hill CH, Zhou A, Bunkóczi G, Stein PE, Marchesan D, Deane JE, Read RJ (2017) Insights into hunter syndrome from the structure of iduronate-2-sulfatase. Nat Commun 8:15786. https://doi.org/10.1038/ncomms15786. PMID: 28593992; PMCID: PMC5472762.

Dinkel C, Drauz K, Gröger H, Schultz C, Waldmann H (2003) Nonindustrial enzyme usage: enzymes in organic synthesis. In: Enzymes in industry: production and applications, pp 263–295. https://doi.org/10.1002/3527602135.ch6a

Eng CM, Guffon N, Wilcox WR, Germain DP, Lee P, Waldek S, Caplan L, Linthorst GE, Desnick RJ (2001) Safety and efficacy of recombinant human α-galactosidase A replacement therapy in Fabry's disease. N Engl J Med 345(1):9–16

Grundtvig IPR, Heintz S, Krühne U, Gernaey KV, Adlercreutz P, Hayler J, Wells A, Woodley JM (2018) Screening of organic solvents for bioprocesses using aqueous-organic two-phase systems. Biotechnol Adv 36(7):1801–1814. https://doi.org/10.1016/j.biotechadv.2018.05.007

Hampe CS, Wesley J, Lund TC, Orchard PJ, Polgreen LE, Eisengart JB, McLoon LK, Cureoglu S, Schachern P, McIvor RS (2021) Mucopolysaccharidosis type I: current treatments, limitations, and prospects for improvement. Biomolecules 11(2):189. https://doi.org/10.3390/biom11020189. PMID: 33572941; PMCID: PMC7911293.

He X, Haselhorst T, von Itzstein M et al (2012) Production of α-L-iduronidase in maize for the potential treatment of a human lysosomal storage disease. Nat Commun 3:1062. https://doi.org/10.1038/ncomms2070

Hout JVD, Reuser A, Klerk JD, Arts W, Smeitink J, Ploeg AVD (2001) Enzyme therapy for pompe disease with recombinant human alpha-glucosidase from rabbit milk. J Inherit Metab Dis 24(2):266–274

Indiani C, Sauve K, Raz A, Abdelhady W, Xiong YQ, Cassino C, Bayer AS, Schuch R (2019) The antistaphylococcal lysin, CF-301, activates key host factors in human blood to potentiate methicillin-resistant *Staphylococcus aureus* bacteriolysis. Antimicrob Agents Chemother 63(4):e02291–e02218. https://doi.org/10.1128/AAC.02291-18. PMID: 30670427; PMCID: PMC6437495.

Jaramillo EH, Sensi A, Brandenberg O, Ghosh K, Sonnino A (2011) Biosafety resource book. FAO, Rome

Jensen B, Eigtved P (1990) Safety aspects of microbial enzyme technology, exemplified by the safety assessment of an immobilized lipase preparation, lipozyme™. Food Biotechnol 4(2):699–725. https://doi.org/10.1080/08905439009549783

Kakkis E (2002) Enzyme replacement therapy for the mucopolysaccharide storage disorders. Expert Opin Investig Drugs 11(5):675–685. https://doi.org/10.1517/13543784.11.5.675

Kakkis E, Muenzer J, Tiller GE, Waber L, Belmont JW, Passage M, Izykowski B, Phillips JD, Doroshow RW, Walot I, Hoft RH, Yu KT, Okazaki S, Lewis DC, Lachman R, Thompson JN, Neufeld EF (2001) Enzyme-replacement therapy in mucopolysaccharidosis I. N Engl J Med 344(3):182–188. https://doi.org/10.1056/nejm200101183440304

Liang X, Wen K, Chen Y, Fang G, Yang S, Li Q (2022) Oral administration of therapeutic enzyme capsule for the management of inflammatory bowel disease. Int J Nanomedicine 17:4843–4860. https://doi.org/10.2147/ijn.s378073

Lu S, Wang X, Lü Q, Hu X, Uppal N, Omenetto FG, Kaplan DL (2009) Stabilization of enzymes in silk films. Am Chem Soc 10(5):1032–1042. https://doi.org/10.1021/bm800956n

Meghwanshi GK, Kaur N, Verma S, Dabi NK, Vashishtha A, Charan PD, Purohit P, Bhandari H, Bhojak N, Kumar R (2020) Enzymes for pharmaceutical and therapeutic applications. Biotechnol Appl Biochem 67(4):586–601. https://doi.org/10.1002/bab.1919

Meyer H, Eichhorn E, Hanlon SP, Lütz S, Schürmann M, Wohlgemuth R, Coppolecchia R (2012) The use of enzymes in organic synthesis and the life sciences: perspectives from the Swiss Industrial Biocatalysis Consortium (SIBC). Catal Sci Technol 3(1):29–40. https://doi.org/10.1039/c2cy20350b

Michaud M, Mauhin W, Belmatoug N, Bedreddine N, Garnotel R, Catros F, Lidove O, Gaches F (2021) Maladie de Fabry: quand y penser? [Fabry disease: a review]. Rev Med Interne 42(2):110–119. https://doi.org/10.1016/j.revmed.2020.08.019. Epub 2020 Nov 7 (French). PMID: 33172708.

Nair SR, Devi CS (2022) Serratiopeptidase: an integrated view of multifaceted therapeutic enzyme. Biomolecules 12(10):1468–1468. https://doi.org/10.3390/biom12101468

Nair NU, Tang WL, Eriksen DT, Zhao H (2014) Industrial applications of enzymes as catalysts. In: Manual of industrial microbiology and biotechnology, pp 480–494. https://doi.org/10.1128/9781555816827.ch33

Porta R, Pandey A, Rosell CM (2010) Enzymes as additives or processing aids in food biotechnology. Enzyme Res 2010:1–2. https://doi.org/10.4061/2010/436859

Raghavendra AJ, Gregory W, Persaud I, Brown JM, Podila R (2018) Spectroscopic insights into the nano-bio interface. In: Unraveling the safety profile of nanoscale particles and materials—from biomedical to environmental applications. IntechOpen. https://doi.org/10.5772/intechopen.69384

Regil RD, Sandoval G (2013) Biocatalysis for biobased chemicals. Biomolecules 3(4):812–847. https://doi.org/10.3390/biom3040812

Richmond JY, McKinney RW (1999) Biosafety in microbiological and biomedical laboratories. US Government Printing Office, Washington, DC. https://ci.nii.ac.jp/ncid/BA26388514.

Schiffmann R, Kopp JB, Austin HA, Sabnis S, Moore DF, Weibel T, Balow JE, Brady RO (2001) Enzyme replacement therapy in Fabry disease a randomized controlled trial. JAMA 285:2743

Sharma S, Mittal A, Gupta VK, Singh H (2006) Improved stabilization of microencapsulated Cathepsin B in harsh conditions. Elsevier BV 40(2):337–342. https://doi.org/10.1016/j.enzmictec.2006.04.024

Silano V, Baviera JMB, Bolognesi C, Cocconcelli PS, Crebelli R, Gott DM, Grob K, Lambré C, Lampi E, Mengelers M, Mortensen A, Rivière G, Steffensen I, Tlustos C, Loveren HV, Vernis L, Zorn H, Herman L, Gomes A et al (2020) Safety evaluation of the food enzyme α-amylase from the genetically modified *Bacillus amyloliquefaciens* strain DP-Czb53. EFSA J 18(10):e06185. https://doi.org/10.2903/j.efsa.2020.6185

Štreimikytė P, Viškelis P, Viškelis J (2022) Enzymes-assisted extraction of plants for sustainable and functional applications. Int J Mol Sci 23(4):2359–2359. https://doi.org/10.3390/ijms23042359

Stütz AE, Wrodnigg TM (2016) Carbohydrate-processing enzymes of the lysosome: diseases caused by misfolded mutants and sugar mimetics as correcting pharmacological chaperones. Adv Carbohydr Chem Biochem 73:225–302

Švarc A, Skendrović D, Presečki AV (2019) Biocatalysis for the production of pharmaceutical intermediates. Bioorg Med Chem 68(9–10):469–476. https://doi.org/10.15255/kui.2019.037

Tao J, Xu J (2009) Biocatalysis in development of green pharmaceutical processes. Curr Opin Chem Biol 13(1):43–50. https://doi.org/10.1016/j.cbpa.2009.01.018

Vanderstocken G, Woolf NL, Trigiante G, Jackson J, McGoldrick R (2022) Harnessing the potential of enzymes as inhaled therapeutics in respiratory tract diseases: a review of the literature. Biomedicines 10(6):1440–1440. https://doi.org/10.3390/biomedicines10061440

Vellard M (2003) The enzyme as drug: application of enzymes as pharmaceuticals. Curr Opin Biotechnol 14(4):444–450. https://doi.org/10.1016/s0958-1669(03)00092-2

Yogalingam G (2004) Aryplase (Biomarin). Curr Opin Investig Drugs (London, England: 2000), 5(10):1111–1120.

Microbial Uricase: A Potential Therapeutic Tool in the Medical Industry

3

Arti Thakur, Nivedita Priya, Mahendra Kumar Varma, and Ajay Kumar

Abstract

Uricase, commonly known as urate oxidase, is an enzyme that catalyzes the conversion of uric acid into allantoin, carbon dioxide, and hydrogen peroxide. In humans, the lack of functional uricase results in the buildup of uric acid, which can lead to hyperuricemia and related health issues such as gout, kidney stones, and tumor lysis syndrome (TLS). A promising approach to address this issue is the utilization of microbial uricase as a therapeutic agent. Microbial sources of uricase offer significant therapeutic potential due to its economical production and versatility, allowing for modifications that can improve its stability and effectiveness in medical applications. This chapter explores the chemistry of uric acid, the diversity of microbial uricase producers, and the intricate pathways involved in uric acid degradation. It also delves into the genetic regulation of uricase expression and the factors influencing its activity. Beyond its biological significance, microbial uricase has far-reaching applications in diagnostics, agri-

A. Thakur (✉)
Department of Allied Health Sciences, Desh Bhagat University, Gobindgarh, Punjab, India
e-mail: Associateprof.AHS@deshbhagatuniversity.in

N. Priya
Department of Medical Laboratory and Life Science, Parul Institute of Allied Health Sciences (PIAHS), Faculty of Medicine, Parul University, Vadodara, Gujarat, India

M. K. Varma
Department of Paramedical Sciences, Vivekananda Global University, Jaipur, Rajesthan, India

A. Kumar (✉)
Department of Industrial Microbiology, Jacob Institute of Biotechnology and Bioengineering, Sam Higginbottom University of Agriculture, Technology and Sciences, Prayagraj, Uttar Pradesh, India
e-mail: ajay.kumar3190@shiats.edu.in

© The Author(s), under exclusive license to Springer Nature Singapore Pte Ltd. 2026

V. Sharma, R. Salwan (eds.), *Pharmaceutically Relevant Microbial Enzymes*,
https://doi.org/10.1007/978-981-95-4941-2_3

culture, and industrial bioremediation. In medicine, it offers innovative solutions for treating hyperuricemia and gout, with new nanozyme-based therapies holding high promise. This chapter underscores the transformative potential of microbial uricase, paving the way for advanced therapeutic applications and sustainable industrial practices.

Keywords

Microbial uricase · Uric acid degradation · Hyperuricemia treatment · Nanozyme-based therapy

3.1 Introduction

The novel therapeutic enzyme uricase, also known as urate oxidase, is a peroxisomal oxidoreductase enzyme that catalyzes the conversion of uric acid into carbon dioxide, hydrogen peroxide, and allantoin—a more soluble form that can be easily eliminated from the body (Roman 2023). This enzymatic reaction is depicted in Fig. 3.1.

This enzyme is absent in humans and certain primates because of an evolutionary deletion of the uricase gene, leading to the buildup of uric acid as the final product of purine metabolism (Roman 2023).

The absence of uricase in humans presents both advantages and disadvantages. On the one hand, uric acid is a potent antioxidant, protecting cells from oxidative stress (Dawson and Walters 2006). On the other hand, excessive accumulation can result in pathological conditions, emphasizing the need for therapeutic interventions. Elevated uric acid levels in the bloodstream, known as hyperuricemia, can lead to debilitating conditions such as gout, kidney stones, and tumor lysis syndrome (TLS) (Richette and Bardin 2010). Effective, targeted therapeutic strategies are crucial for managing these diseases. Uricase, a highly selective enzyme for uric acid, emerges as a promising solution. This enzyme is essential in medical applications, from detecting uric acid in blood and urine to being used as a protein-based treatment to reduce the harmful accumulation of urates, offering significant potential for improving patient outcomes.

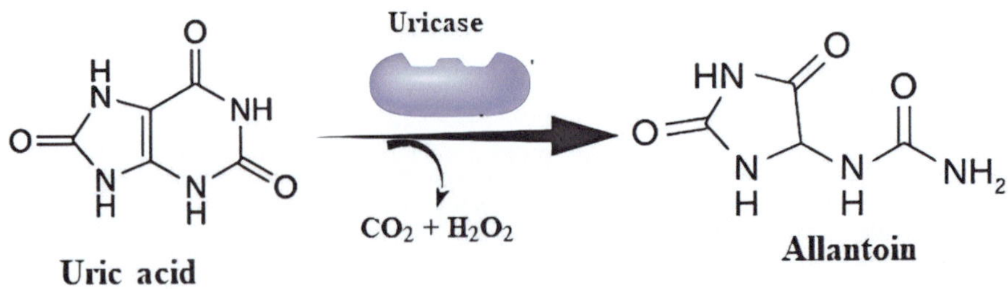

Fig. 3.1 Illustration of uricase-mediated breakdown of uric acid

Microbes, animals, and plants can produce uricase enzyme. However, microbes provide the advantages of faster growth rates, cost-effective bioprocessing, and easier optimization of the growth medium (Wan et al. 2023). Advances in biotechnology have enabled the production of recombinant uricase, which has shown efficacy in lowering uric acid levels in patients with refractory gout and other uric acid-related disorders (Wu et al. 1994; Richette and Bardin 2010). This chapter explores the chemical nature of uric acid, its biological significance, degradation, and the potential of microbial uricase as a therapeutic agent. This chapter explores the therapeutic potential of microbial uricase, its biochemical properties, production, clinical applications, and future prospects.

3.1.1 Uric Acid: Chemical Structure and Biosynthesis

Uric acid is a heterocyclic compound made up of carbon, nitrogen, oxygen, and hydrogen with the formula $C_5H_4N_4O_3$. Its structure includes a purine backbone with a hydroxyl group at position eight and two keto groups at locations two and six (Fig. 3.2).

These functional groups contribute to the acidic characteristics, with pK_a values ranging between 5.4 and 10.3 (Mandal et al. 2022). It forms different salts and ions such as acid urates, urates, and ammonium acid urates (Ridi et al. 2017).

Uric acid, a weak diprotic acid with pK_a values of 5.4 and 9.8, mainly exists as the deprotonated urate anion at physiological pH as depicted in Fig. 3.3. For males and premenopausal females, the reference ranges for sUA are 3.5–7.2 mg/dL (210–430 μmol/L) and 2.6–6.0 mg/dL (155–360 μmol/L), respectively (Desideri et al. 2014).

In 1776, Swedish chemist Scheele made a groundbreaking discovery by identifying uric acid as an essential component of bladder stones (Chen et al. 2016). Uric acid is biosynthesized in the liver as the end product of purine metabolism. Uric acid is generated through the metabolism of both endogenous purines, which are synthesized at daily rates of approximately 300–400 mg, and exogenous purines,

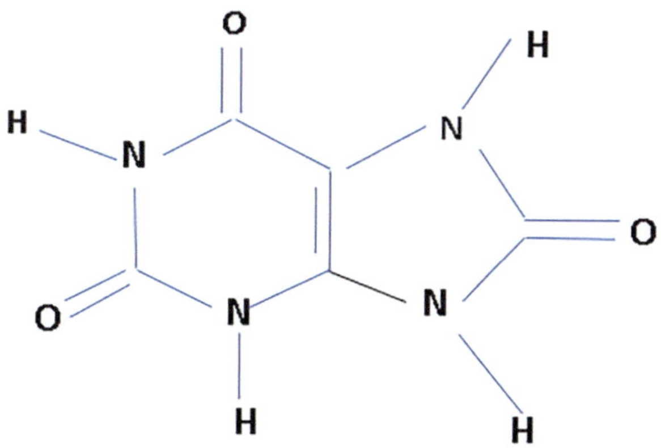

Fig. 3.2 Structure of uric acid

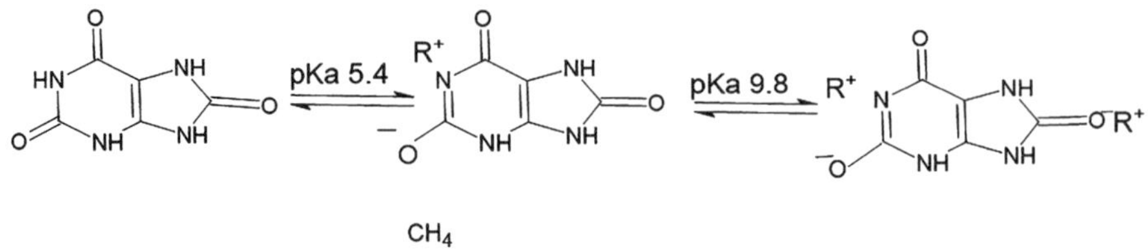

Fig. 3.3 Structure of uric acid, pK_a, urate salt, and the process of salt formation

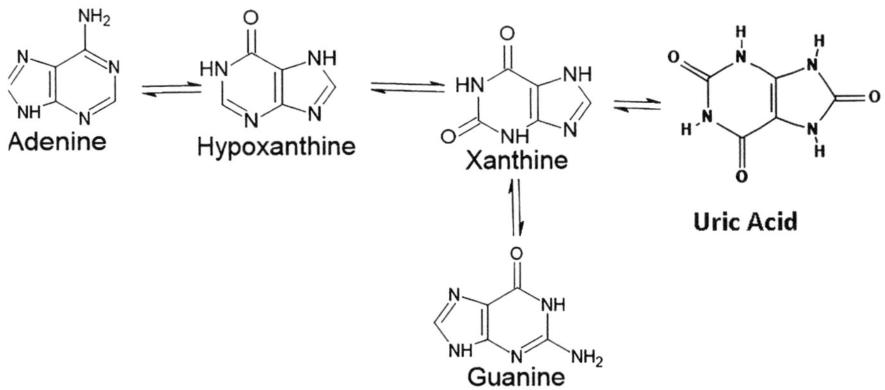

Fig. 3.4 Illustration of uric acid production by the metabolic conversion of purines (adenine and guanine)

derived from dietary sources, contributing around 300 mg. In healthy males, the total uric acid pool is approximately 1200 mg, while in females, it is about 600 mg, particularly when following a purine-free diet (Loffler et al. 1982; Maiuolo et al. 2016).

The enzyme xanthine oxidase catalyzes the oxidation of hypoxanthine to xanthine and then xanthine to uric acid as shown in Fig. 3.4 (Desai et al. 2023).

In most mammals, uricase converts uric acid into allantoin, a highly water-soluble molecule, making it easier for the body to eliminate. However, in humans, the lack of uricase leads to the direct excretion of uric acid via the kidneys (Dawson and Walters 2006). The absence of uricase activity in humans is primarily linked to a nonsense mutation at codon 33 in exon 2, another nonsense mutation at codon 187, and a splice mutation in exon 3 (Wu et al. 1992). The independent mutations that arose during the evolution of hominoids, alongside the evolution of both Old World and New World monkeys, may suggest that early primates experienced a natural selective advantage associated with higher levels of uric acid (Watanabe et al. 2002).

3.1.2 Benefits and Hazards of Uric Acid

Benefits

Uric acid functions as a significant antioxidant in the blood, scavenging free radicals and providing protection against oxidative damage (Dawson and Walters 2006). It has been implicated in cognitive enhancement, potentially offering a survival advantage in early human evolution (Wu et al. 1994; Lu et al. 2016). It has also been demonstrated to be essential for blood pressure maintenance (Watanabe et al. 2002) and to offer physiological protection against cancer and aging (Ames et al. 1981).

Hazards

Hyperuricemia is a main risk factor for gout, a disorder defined by the deposition of urate crystals in joints, producing inflammation and pain (Richette and Bardin 2010). Elevated uric acid levels are associated with an increased risk of metabolic syndrome, cardiovascular diseases, chronic kidney disease, and liver illness (Dawson and Walters 2006). Hypouricemia, or excessively low UA levels, has been linked to weakened physiological defenses (such as antioxidant properties) (Wen et al. 2023). Autoimmune conditions, including lichen planus, Parkinson's disease, and pemphigus vulgaris, are associated with reduced uric acid levels (Bakhtiari et al. 2017; Kuwabara et al. 2023). Elevated uric acid levels may trigger the release of certain autacoids and hormones, increasing blood pressure. Additionally, the infiltration of uric acid into the smooth muscle of the heart can contribute to a range of cardiovascular diseases (Kanbay et al. 2013; Kuwabara et al. 2023). Through the lens of its dual role, uric acid is a key molecule, emphasizing the therapeutic value of uricase in maintaining metabolic equilibrium. A summary of the factors that may affect serum urate levels and their implications for human health is presented in Fig. 3.5.

3.2 Uric Acid-Degrading Microorganisms

A diverse range of microorganisms possess the ability to break down uric acid via some enzymes like uricase (urate oxidase). These microbes thrive in a variety of environments and are essential for converting uric acid into simpler, more soluble compounds such as allantoin, urea, ammonia, and carbon dioxide. The first microbial uricase was extracted from the fungus *Neurospora crassa* in 1957 (Perez-Ruiz et al. 2014). Subsequently, it was found that several bacterial species are capable of degrading uric acid, including *Pseudomonas aeruginosa*, *Bacillus thermocatenulatus*, *Microbacterium* sp., *Arthrobacter globiformis*, *Nocardia farcinica*, *Escherichia coli*, *Bacillus subtilis*, and *Bacillus fastidious* (Saeed et al. 2004; Suzuki et al. 2004; Zhou et al. 2005; Lotfy 2008; Xu et al. 2022; Chen and Li 2023). The uricase enzyme is produced extracellularly in *Aspergillus terreus*, *Penicillium chrysogenum*, *Pseudomonas aeruginosa*, *Bacillus subtilis*, *Escherichia coli*, *Alcaligenes*, *Bacillus thermocatenulatus*, and *Streptomyces* sp. On the other hand, it is produced intracellularly in *Micrococcus luteus*, *Candida utilis*, *Klebsiella pneumoniae*,

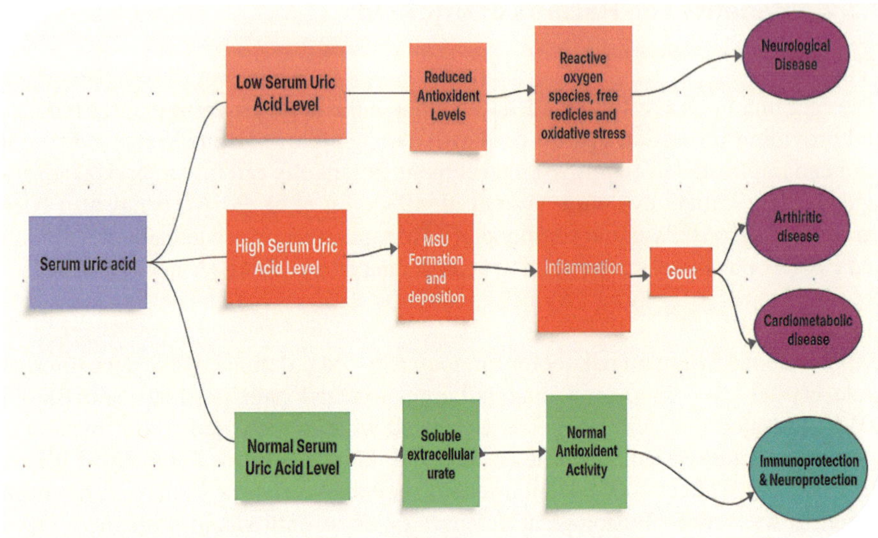

Fig. 3.5 Various determinants of serum uric acid levels and their human diseases. SUA serum uric acid, ROS reactive oxygen species, MSU monosodium urate

Table 3.1 Summary of extracellular versus intracellular uricase producers

Type	Microorganisms	References
Extracellular	*Aspergillus terreus*, *Penicillium chrysogenum*, *Pseudomonas aeruginosa*, *Bacillus subtilis*, *Escherichia coli*, *Alcaligenes*, *Bacillus thermocatenulatus*, and *Streptomyces* sp.	Atalla and El-Shamy (2013), Abdel-Fattah et al. (2005), Pustake et al. (2019)
Intracellular	*Micrococcus luteus*, *Candida utilis*, *Klebsiella pneumoniae*, *Microbacterium*, *Proteus vulgaris*, *Streptomyces albidoflavus*, *Streptomyces graminofaciens*, *Pseudomonas aeruginosa*, and *Streptomyces exofolitus*	Yoshimoto et al. (1991), Zhou et al. (2005), Kim et al. (2004), Suh et al. (2005), Abdel-Fattah et al. (2005), Pustake et al. (2019), Parvez et al. (2016), Chen and Li (2023)

Microbacterium, *Proteus vulgaris*, *Streptomyces albidoflavus*, *Streptomyces graminofaciens*, *Pseudomonas aeruginosa*, and *Streptomyces exofolitus* (Yoshimoto et al. 1991; Zhou et al. 2005; Kim et al. 2004; Suh et al. 2005; Abdel-Fattah et al. 2005; Atalla and El-Shamy (2013); Parvez et al. 2016; Chen and Li 2023) (Table 3.1).

Some other bacterial species of lactic acid bacteria, such as *Pediococcus* species, *Bifidobacterium*, *Leuconostoc*, *Lactobacillus*, and *Enterococcus*, have also been reported to degrade uric acid. In an experiment, administration of these bacterial species in rats/mice led to the reduction in serum uric acid level (Ogawa 2006; Li et al. 2023; Negm et al. 2023), suggesting their potential role in modulating uric acid levels in vivo. Different soil fungi, including *Fusarium*, *Helminthosporium*,

Spondilocladium, *Curvularia*, *Stemphylium*, *Aspergillus*, *Geotri chum*, *Penicillium*, *Mucor*, *Rhizopus*, *Alternaria*, and *Chaetomium*, were also reported to be capable of producing intracellular uricase at high rates, particularly in the presence of urea or uric acid as the sole nitrogen source (Geweely and Nawar 2011; Aly et al. 2013; Rajagopalan et al. 2017).

3.3 Uric Acid Degradation Pathway in Bacteria

The uric acid degradation in bacteria encompasses enzymatic pathways that convert uric acid (2,6,8-trihydroxypurine) into simpler, more utilizable compounds. Uric acid is an important nitrogenous compound that acts as a metabolic substrate for various bacterial species. The capacity of bacteria to metabolize uric acid plays a crucial role in the recycling of nitrogen within ecosystems, allowing these microorganisms to flourish in a wide range of habitats. Bacteria employ several enzymatic pathways to transform uric acid into more basic substances, including allantoin, ammonia, and glyoxylate. The purine catabolism to uric acid process is almost conserved across various microorganisms and is shown in Fig. 3.6.

However, the final product resulting from uric acid degradation differs among species, contingent upon the specific active catabolic enzymes present. Xanthine emerges as the initial intermediate product in the degradation pathway of all purine bases. Within the cytosol, xanthine dehydrogenase catalyzes the breakdown of xanthine, resulting in the production of urate. This compound is subsequently transported into the peroxisome, which undergoes oxidation mediated by uricase, forming 5-hydroxyisourate. The evolution of different urate oxidase, which includes allantoicase, uricase, and allantoinase, may account for the different degradation

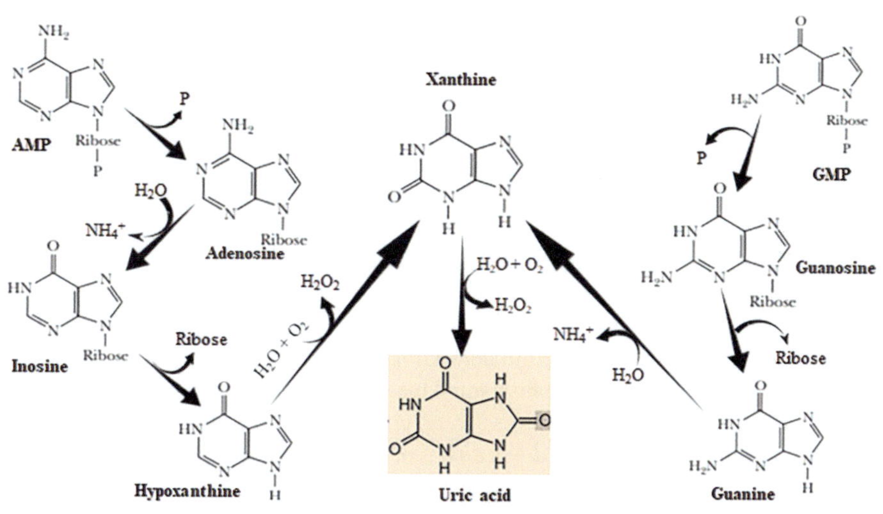

Fig. 3.6 Illustration of enzymatic degradation of purines to uric acid

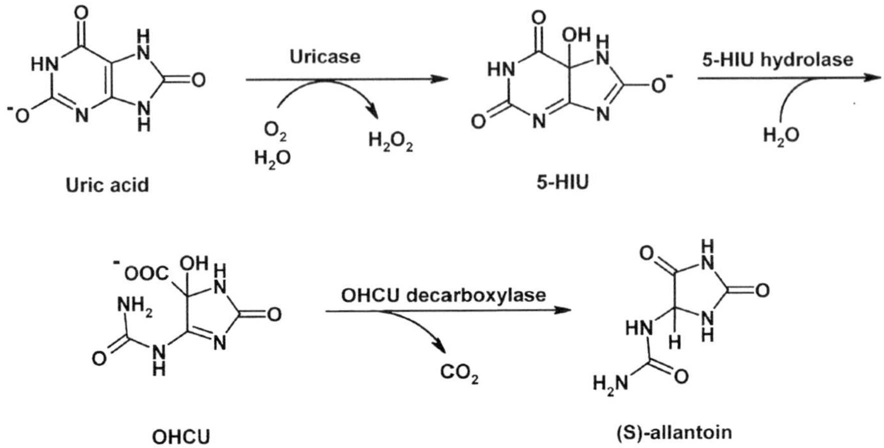

Fig. 3.7 Schematic representation of uric acid degradation pathway

by-products of uric acid observed in various microorganisms. Some of the metabolic pathways explaining these variations are discussed below.

3.3.1 The Uricase Pathway

The uricase pathway represents the predominant mechanism for the degradation of uric acid in bacterial organisms, and the pathway is illustrated in Fig. 3.7. The process begins with uricase, which catalyzes the oxidation of uric acid to form 5-hydroxyisourate (5-HIU). This intermediate compound is subsequently hydrolyzed by 5-HIU hydrolase, producing 2-Oxo-4-hydroxy-4-carboxy-5-ureidoimidazoline (OHCU). A third enzyme, known as OHCU decarboxylase, facilitates the decarboxylation of OHCU, resulting in the production of (S)-allantoin (Vogels and Van der Drift 1976; Lu et al. 2023). *Bacillus paramycoides* YC02 & *Bacillus fastidiosus* strains have been identified to degrade uric acid via this pathway (Bongaerts and Vogels 1976).

3.3.2 The Allantoin Degradation Pathway

Allantoin, the final product of the uricase pathway, undergoes further degradation into urea and glyoxylate in certain bacterial species (Fig. 3.8). Allantoin is a common by-product of purine metabolism. Due to its high nitrogen content, many organisms can degrade it and repurpose its components. Certain bacteria, such as *Bacillus fastidiosus, Delftia acidovorans, and Cupriavidus pinatubonensis* JMP13 can degrade allantoin in the presence of oxygen, whereas others, including *Escherichia coli* K-12, *Streptococcus allantoicus*, and *Providencia rettgeri*, can

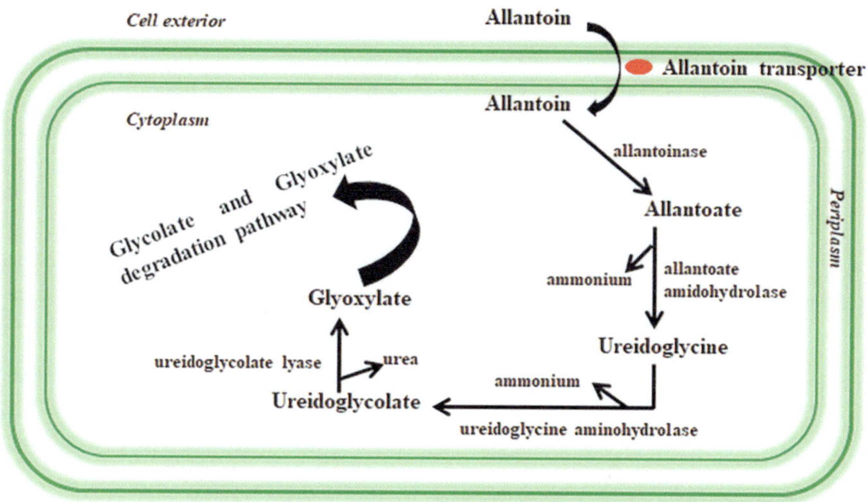

Fig. 3.8 Illustration of the allantoin pathway of uric acid degradation

degrade it only under anaerobic conditions (Kim et al. 2021). The initial phase of allantoin degradation is facilitated by allantoinase, an enzyme that hydrolyzes the internal amide bond of (S)-(+)-allantoin, resulting in the formation of allantoate (Vogels and Van der Drift 1976). This enzyme is found universally in all organisms that can metabolize allantoin and represents the first step in all recognized allantoin degradation pathways.

Allantoate undergoes conversion to (S)-ureidoglycolate via one of the two distinct mechanisms. In the first pathway, nitrogen is liberated in a single step as urea, a reaction facilitated by the enzyme allantoiase (Lee et al. 2013). In contrast, the second pathway involves a two-step process where nitrogen is released as ammonia, passing through the intermediate (S)-Ureidoglycine, with the initial step catalyzed by ureidoglycine aminohydrolase, which is reported to be found in many bacterial species, such as in *E. coli* O157:H7 (Vogels and Van der Drift 1976).

3.3.3 Allantoin Fermentation

Allantoin fermentation represents a metabolic pathway employed by specific bacteria to anaerobically break down uric acid, allowing these organisms to utilize allantoin as a nitrogen source in environments with limited oxygen availability. This process is essential in oxygen-deprived settings and encompasses a sequence of enzymatic reactions that transform uric acid into energy and nitrogenous substances as illustrated in Fig. 3.9. Some bacterial species that demonstrate this remarkable biochemical versatility include *Streptococcus allantoicus* and *Streptococcus allantoic* (Valentine et al. 1962), *Escherichia coli*, *Streptomyces coelicolor*, and *Bacillus subtilis* (Smith et al. 2021) etc.

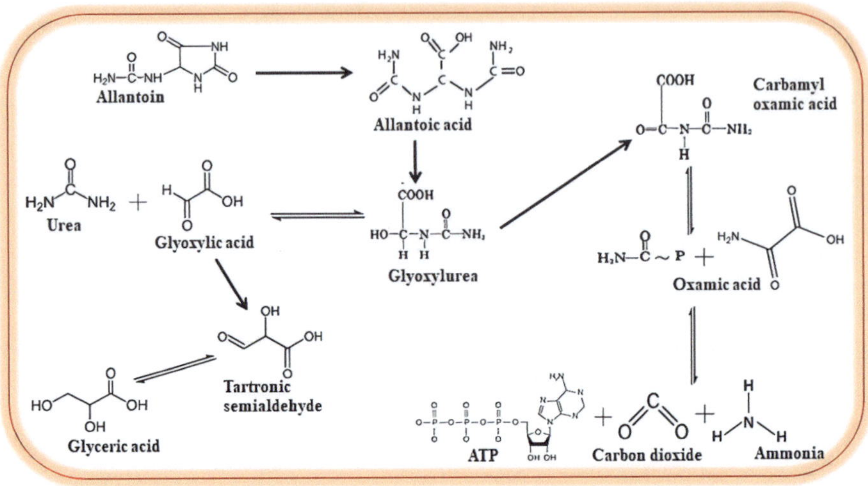

Fig. 3.9 A schematic presentation of allantoin fermentation pathway of bacterial uric acid degradation

Mechanism of Allantoin Fermentation

Uric Acid Reduction: In the absence of oxygen, bacteria break down uric acid by converting it into xanthine, facilitated by the enzyme urate oxidoreductase (Vogels and Van der Drift 1976). Xanthine undergoes further reduction to form hypoxanthine. Hypoxanthine is converted through deamination into xanthine and ammonia (Vogels and Van der Drift 1976). Xanthine undergoes conversion into uric acid, which is subsequently transformed into allantoin. Allantoin is converted into allantoate through the action of the enzyme allantoinase (Ramazzina et al. 2006). Allantoate undergoes additional degradation to form ureidoglycolate and ammonia. Ureidoglycolate is transformed into glyoxylate along with the release of extra ammonia (Ramazzina et al. 2006). Further, ureidoglycolate is converted into oxamic acid and glyceric acid (Lee et al. 2013).

The limited similarity between the aerobic and anaerobic pathways of allantoin degradation was demonstrated by Barker (1943), who identified oxamic acid as a primary product resulting from the fermentation of allantoin by *Streptococcus allantoicus*.

Formate-Dependent Uric Acid Degradation Pathway

Formate-dependent uric acid degradation also represents an anaerobic metabolic pathway employed by specific bacteria to anaerobically break down uric acid. This pathway is particularly significant in environments like the human gut, where bacteria use formate as an electron donor to degrade uric acid (Li et al. 2017; Wei et al. 2022a). This pathway unfolds through a series of intricately coordinated stages, each playing a pivotal role in optimizing metabolic efficiency under anaerobic conditions. The process initiates with the oxidation of formate to carbon dioxide (CO_2), leading to the liberation of electrons (Moreno et al. 2023a). The electrons generated from the oxidation of formate are later conveyed to the electron transport chain, where they aid in the reduction of uric acid (Barbieri et al. 2020a). Uric acid is transformed into xanthine and several other metabolites, contributing to nitrogen recycling and participating in the organism's energy production (Wei et al. 2022b). The formate-dependent degradation pathway is particularly prominent within the human gut microbiota, contributing significantly to the regulation of uric acid levels (Smith et al. 2021). Identification of key enzymes: Studies have elucidated the role of enzymes such as formate dehydrogenase and urate reductase, which are essential for anaerobic uric acid degradation (Vidal et al. 2022a; Wang et al. 2021).

3.4 Uric Acid-Degrading Enzymes

Uric acid-degrading enzymes or uricase enzymes, which are proteins, exhibit characteristics typical of cofactors and play a crucial role in catalyzing a variety of biochemical reactions. A deficiency in these enzymes may result in the development of various diseases. Uricase is frequently utilized in biochemical diagnostics, particularly for the detection of blood uric acid levels, as well as in industrial applications (Ridi et al. 2017; Tandon et al. 2021). Therapeutic enzymes are distinct from conventional pharmaceuticals due to their interactions with pathogens, which may involve the activation or inhibition of specific reactions and a direct connection to the substrate responsible for the disease through its accumulation in the body (Meletis and Barker 2005). These therapeutic enzymes can be obtained biologically from sources such as fungi, plants, and bacteria, or they can be synthesized in laboratory settings. Uricase is composed of four identical subunits arranged in a tetrameric structure. Each monomer consists of two distinct domains of type 2 copper binding sites that allow it to bind and convert uric acid into hydrogen peroxide and allantoin (Wu et al. 1989). Uric acid degradation is facilitated by a range of specialized enzymes that transform uric acid into more soluble and excretable substances. Below is a comprehensive list of some common uric acid-degrading enzymes, arranged by type of organism and supported by thorough descriptions, roles, and current scholarly references.

Bacteria are highly efficient in metabolizing uric acid, often using it as a carbon and nitrogen source. The enzymatic machinery in bacteria for uric acid degradation is highly conserved and widely studied. The following enzymes are found in bacteria along with their roles: This enzyme catalyzes the oxidation of uric acid into 5-hydroxyisourate and plays a central role in uric acid metabolism under aerobic conditions. It has been reported in organisms such as *Bacillus subtilis*, *Escherichia coli*, and *Pseudomonas aeruginosa* (Moreno et al. 2023b). This enzyme converts 5-hydroxyisourate into 2-oxo-4-hydroxy-4-carboxy-5-ureidoimidazoline (OHCU) and is essential for the second step of uric acid degradation, ensuring the breakdown of intermediates into non-toxic compounds. It has been detected in *Klebsiella pneumoniae* and *Pseudomonas putida* (French and Ealick 2011). This enzyme decarboxylates OHCU to produce allantoin and supports nitrogen recycling in bacterial ecosystems. It has been reported in Gram-negative bacteria such as *Escherichia coli* (Vidal et al. 2022b).

This enzyme hydrolyzes allantoin to allantoic acid and is a key enzyme in nitrogen metabolism, especially in nitrogen-limited environments. It has been commonly reported in *Clostridium* species and other anaerobic bacteria (Wei et al. 2022c). This enzyme breaks down ureidoglycolate into glyoxylate and ammonia in the final step of uric acid degradation, completing the nitrogen recycling pathway. It has been found in *Proteobacteria* and *Actinobacteria* (Li et al. 2017). Fungi exhibit unique uric acid degradation pathways, often facilitating nutrient recycling in their environments. Many enzymes play a vital role in this context. This enzyme oxidizes uric acid into 5-hydroxyisourate. It has been found in *Aspergillus nidulans*, *Candida albicans*, and *Neurospora crassa*. Urate oxidase plays a crucial role in fungal growth, particularly in nitrogen-limited environments (Barbieri et al. 2020b). This enzyme converts 5-hydroxyisourate into OHCU and plays a significant role in reducing the accumulation of toxic intermediates during uric acid breakdown. It has been detected in *Penicillium chrysogenum* and *Neurospora* species (Smith et al. 2021). This enzyme hydrolyzes allantoin to produce allantoic acid and supports nutrient recycling in decomposer fungi. It has been found in various filamentous fungi (Vidal et al. 2022b). Archaea exhibit distinctive adaptations for the metabolism of uric acid, especially in extreme environmental conditions. The following enzyme has been reported in Archaea: This enzyme catalyzes the oxidation of uric acid under extreme environmental conditions, facilitating nitrogen recycling in nutrient-limited and hypersaline environments. It has been found in halophilic archaea such as *Halobacterium salinarum* (Li et al. 2021).

In animals, uric acid degradation is limited due to the absence of uricase in many species, including humans. However, some enzymes are still functional in uric acid metabolism which converts uric acid to allantoin. It is functional in non-primate mammals such as rodents and canines. It is critical for maintaining low serum uric acid levels (Smith et al. 2021). This enzyme hydrolyzes allantoin to allantoic acid and is present in amphibians and reptiles. It plays a crucial role in nitrogen waste management in these organisms (Barbieri et al. 2020b). This enzyme converts hypoxanthine to xanthine and xanthine to uric acid, indirectly influencing uric acid levels and purine metabolism. It has been found in humans, cattle, and other mammals (Vidal et al. 2022b).

Plants utilize uric acid-degrading enzymes for nitrogen recycling and reutilization. This enzyme facilitates the conversion of uric acid into allantoin and serves as a crucial element in nitrogen metabolism and stress response mechanisms in plants. It has been identified in species such as *Arabidopsis thaliana*, *Glycine max* (soybean), and *Medicago truncatula* (Jiang et al. 2022). This enzyme catalyzes the hydrolysis of allantoin to allantoic acid and is involved in nitrogen fixation and storage processes. It is found in legumes, including *Phaseolus vulgaris* (common bean) (Wei et al. 2022c).

This enzyme transforms allantoic acid into ureidoglycolate, thereby facilitating nitrogen recycling in environments with limited nutrients (Li et al. 2017). Formate dehydrogenase plays a vital role in the uric acid degradation pathway by facilitating the oxidation of formate, which is generated as a by-product during urate degradation, into carbon dioxide. The enzyme utilizes cofactors like NAD+ or cytochrome c to accept electrons during the oxidation process, which helps generate reducing power or facilitate electron transport for cellular respiration (Wei et al. 2022b; Vidal et al. 2022b). This process is critical for energy conservation and the recycling of carbon in various organisms. It was reported in bacteria such as *Escherichia coli*, *Pseudomonas aeruginosa*, and *Clostridium* species. It is also present in fungi and certain plants (Moreno et al. 2023b).

Urate Reductase

Urate reductase facilitates the conversion of uric acid into 5,6-dihydroxyuric acid, which serves as an intermediate in the anaerobic degradation pathway of uric acid. This enzyme functions under reducing conditions and is crucial for the survival of anaerobic bacteria and fungi (Li et al. 2021). It has been identified in species such as *Clostridium acetobutylicum*, *Methanobrevibacter smithii* (archaea), and various anaerobic fungi (Jiang et al. 2022). Additionally, urate reductase is being investigated for its potential in enzyme-based therapies aimed at lowering serum uric acid levels in patients suffering from hyperuricemia and gout (Barbieri et al. 2020a).

3.5 Genetics and Uricase-Encoding Genes

The process of bacterial degradation of uric acid encompasses a sequence of enzymatic reactions, each facilitated by particular enzymes encoded by unique genes. Fourteen functional genes have been identified that encode enzymes and proteins involved in the purine catabolic pathway. The activity of xanthine dehydrogenase necessitates the expression of five specific genes: *pucE*, *pucD*, *pucC*, *pucB*, and *pucA*. Additionally, two genes, *pucM* and *pucL*, are responsible for encoding uricase, while two other genes, *pucK* and *pucJ*, are involved in the uric acid transport system. The genes *pucI*, *pucH*, and *pucF* encode for allantoin permease, allantoinase, and allantoate amidohydrolase, respectively. Research has shown that the *pucR*-mutant strain of *Bacillus subtilis* exhibited the lowest activity among all the genes tested, suggesting that PucR plays a regulatory role in the expression of the *puc* genes. All 14 genes except *pucI* are located at 284°–285° in the gene cluster on the chromosome and are implicated in six transcription units. Allantoic acid, allantoin, and uric acid act as effector compounds to regulate PucR for the expression of *puc* genes. These genes are frequently arranged in clusters within bacterial genomes, which allows for the coordinated regulation of uric acid degradation. For example, a novel bacterium strain 44XB^{T} revealed the presence of multiple copies of the genes *pucL*, *pucM*, *uraD*, and *allB*, including six tandem copies of *pucL* that exhibited an amino acid identity ranging from 70.42% to 99.70% (Liu et al. 2024).

The expression of the *puc* genes is regulated by uric acid, allantoic acid, and allantoin compounds through PucR (Argyrou et al. 2001). Uricases, found in aquatic vertebrates and microbes, are predominantly soluble and located in the cytoplasm of bacteria or the peroxisomes of yeasts (Kratzer et al. 2014). The catabolic genes associated with uric acid in *C. neoformans* have been identified as URO1 (urate oxidase), URO2 (HIU hydrolase), URO3 (OHCU decarboxylase), DAL1 (allantoinase), DAL2,3,3 (a fusion protein of allantoicase and ureidoglycolate hydrolase), and URE1 (urease) (Lee et al. 2013). Uricases of most microbial and aquatic vertebrate species are soluble and remain in either the cytoplasm (bacteria) or peroxisome (yeast) (Kratzer et al. 2014).

Nonsense mutations have led to the pseudogenization of the uricase gene in humans. Although this gene is non-functional, cDNA sequencing has confirmed the presence of uricase mRNA in human liver cells, which contains two premature stop codons (Wu et al. 1989; Yeldandi et al. 1991).

Bacteria and fungi possess the ability to utilize a wide range of compounds, including purines, for nitrogen and carbon sources. In the case of *Pseudomonas aeruginosa*, the genes responsible for the initial deamination of adenine and guanine, which serve as nitrogen sources, are situated at distinct loci on the chromosome. Conversely, the genes that encode the enzymes responsible for the degradation of hypoxanthine to ureidoglycolic acid are found near one another (Matsumoto et al. 1978). It has been recently reported that three bacterial species, *C. sporogenes*, *L. saccharolytica*, and *C. aerofaciens*, have been identified as sharing five uric-acid genes that exhibit the highest levels of induction: *ygeX*, *ygeY*, *ygeW*, *ygfK*, and *ssnA*, determined by RNA sequence (RNA-seq) analysis. The putative functions of these gene products include ammonia lyase (YgeX), peptidase (YgeY) (Liu et al. 2023a), carbamoyl transferase (YgeW), oxidoreductase (YgfK), and amidohydrolase (SsnA). These enzymes are believed to play a role in the reduction and cleavage of bonds found in uric acid.

3.5.1 Genetic Organization of Uric Acid Degradation Pathway

The genes responsible for uric acid degradation are evolutionarily conserved across various taxa, with specific adaptations in bacteria, fungi, and archaea. Table 3.2. summarizes the key genes involved in the pathway.

3.5.2 Pathway Regulation and Genetic Architecture

Bacterial Regulation: The hpx Operon

In *Escherichia coli* and other bacteria, uric acid degradation genes are part of the **hpx operon**, regulated by the **PurR** repressor (Kim et al. 2000). When purines are abundant, PurR represses the operon. Under nitrogen-limited conditions, these genes are activated to facilitate nitrogen scavenging (Kim et al. 2000).

Fungal Regulation and Evolutionary Adaptation

In fungi like *Aspergillus nidulans* and *Neurospora crassa*, these genes are upregulated under nitrogen starvation (Barbieri et al. 2020b). The evolution of *uox* in fungi

Table 3.2 Genes involved in uric acid degradation in microorganisms

Reaction step	Enzyme	Gene name	Function	Microbial groups	Reference
Step 1: Uric acid oxidation	Urate oxidase (Uricase)	*uox*, *hpxO*	Converts uric acid into 5-hydroxyisourate (5-HIU)	Bacteria (*Bacillus subtilis*), fungi (*Aspergillus nidulans*), Archaea (*Halobacterium salinarum*)	Vogels and Van der Drift (1976), Kahn et al. (2021)
Step 2: Hydrolysis of 5-HIU	5-Hydroxyisourate hydrolase	*hiuH*, *hpxT*	Hydrolyzes 5-hydroxyisourate to OHCU	Bacteria (*Pseudomonas putida*), fungi (*Neurospora crassa*)	Vogels and Van der Drift (1976), Barbieri et al. (2020a)
Step 3: Decarboxylation of OHCU	OHCU decarboxylase	*pucL*, *hpxD*	Converts OHCU into allantoin	Bacteria (*Klebsiella pneumoniae*), fungi (*Penicillium chrysogenum*)	Vidal et al. (2022a)
Step 4: Hydrolysis of allantoin	Allantoinase	*allB*, *pucA*	Hydrolyzes allantoin to allantoic acid	Bacteria (*Escherichia coli*, *Clostridium* species), fungi	Wei et al. (2022c)
Step 5: Conversion of allantoic acid	Allantoicase	*allC*, *pucB*	Converts allantoic acid into ureidoglycolate	Bacteria (*Pseudomonas aeruginosa*), plants (*Zea mays*)	Jiang et al. (2022)
Step 6: Final breakdown step	Ureidoglycolate hydrolase	*pucM*, *uglH*	Converts ureidoglycolate into glyoxylate and ammonia	Bacteria (*Actinobacteria*, *Proteobacteria*)	Li et al. (2017)

suggests adaptation to nitrogen-poor environments, particularly in decomposer fungi (Wei et al. 2022b).

Uric Acid Metabolism in Archaea

In extremophiles like *Halobacterium salinarum*, **uox** functions under hypersaline conditions, ensuring nitrogen conservation (Nguyen et al. 2023).

Horizontal gene transfer has played a significant role in adapting these genes to extreme environments (Vidal et al. 2022a).

3.6 Uricase Activity Affecting Factors

In microbial systems, the functionality of uricase is affected by various factors, such as environmental conditions, the availability of substrates, and genetic regulation. The following outlines the primary factors that influence the activity of microbial uricase.

Uricase activity is highly sensitive to changes in pH levels. Most microbial uricases function optimally within an alkaline range, typically between pH 7.5 and 9.0. When exposed to excessively acidic or basic conditions, the enzyme may lose its stability, leading to denaturation or reduced catalytic efficiency. Temperature is another crucial factor influencing uricase activity. Most microbial uricases exhibit peak performance within the temperature range of 30–50 °C. Outside this range, enzyme activity tends to decline, and prolonged exposure to extreme temperatures can result in denaturation. Research findings highlight variations in optimal growth parameters for different microbial strains. Manimekalai and Anandharaj (2021) determined that *Bacillus subtilis* MM13 produces uricase most effectively at pH 7.0 and temperature 40 °C. Similarly, Atalla et al. (2009) observed maximum uricase activity at pH 8.0, and temperature 30 °C, with an incubation period of 8 days and an aeration level of 150 rpm.

Further studies reinforce these observations. Ravichandran et al. (2015) reported an optimal pH of 8 and a temperature range of 25–45 °C for uricase activity. However, they found that exposure to 60 °C for 1 h led to a 50% decline in enzymatic activity. In contrast, Geweely and Nawar (2011) noted that *Aspergillus niger*-derived uricase functions best at 35 °C. Apart from pH and temperature, metal ions also play a role in uricase activity. While certain heavy metals act as cofactors for enzymatic reactions, excessive concentrations can have an inhibitory effect (Suzuki et al. 2004). The optimum temperature for uricase activity (0.261 U/ml) in *Halobacillus* sp. strain GCFx14 was identified as 40 °C. This strain produced uricase with an activity of 0.233 U/ml at pH 9, while a 1% salinity level resulted in 0.215 U/ml activity. The highest uricase production was achieved under pH 8.0, 35 °C, 2.5% salinity, and 0.2% uric acid. Additionally, yeast extract and peptone (meat) were found to be the most effective nitrogen and carbon sources for inducing uricase production in this strain.

These studies collectively emphasize the importance of environmental and nutritional factors in optimizing microbial uricase activity, which has significant implications for industrial and medical applications. Heavy metals act as cofactors for certain enzymes; however, excessive amounts may hinder enzyme activities (Suzuki et al. 2004). On the contrary, Ravichandran et al. (2015) observed a 140% increase in enzyme activity due to copper stimulation. Consequently, unidentified elements may play a role in the activation and inhibition of enzymes, with variations observed across different bacterial species (Ravichandran et al. 2015).

3.7 Applications of Uricase

Uricase is widely distributed in soil as well as in aquatic environments, where it plays an essential role in nitrogen metabolism as illustrated in Fig. 3.10. It is synthesized by certain microbes like yeast, bacteria, fungi, and algae, and has been of great research interest owing to the potential use in waste treatment, fertilizer and soil conditioning, animal feed, and drugs (Sumner and Somers 1947; Varner 1959). Microbial enzyme uricase is of great importance in industrial and agricultural uses due to its wide range of applications. Uricase has great industrial applications, especially in food and beverages. It is utilized as a measuring tool for the quantity of uric acid in beer and wine during fermentation. Any deviation from the normal quantity of uric acid could be a sign of probable contamination or spoilage. Uricase is also utilized in the production of allantoin, which has its uses in various functions in the pharmaceutical and cosmetic industries. Allantoin has uses in moisturizing the skin and anti-inflammatory purposes and is thus a highly demanded ingredient incorporated into creams, lotions, and other cosmetic products. Besides that, microbial uricase is vital in the nitrogen cycling process since it oxidizes uric acid present in animal manure into more beneficial nitrogenous compounds and thus increases soil fertility. Some of its applications in agriculture are as follows.

3.7.1 Agricultural Applications

Soil Fertility and Nutrient Recycling
Uric acid is a very important nitrogenous product in animal dung. The buildup of uric acid in the soil could lead to nitrogen imbalances, which could lower the fertility of the soil. The microbial enzyme uricase helps in the breakdown of uric acid into easily utilizable nitrogen products, including ammonia and urea, thus improving the health of the soil and plant growth (Liu et al. 2023a).

Decomposition of Animal Manure and Increased Crop Yields Animal dung, particularly poultry dung, is characterized by high levels of uric acid. Microbial uricase-producing bacteria like *Bacillus subtilis* and *Aspergillus flavus* can be applied to composting processes to help break down uric acid. The process not only helps in the reduction of environmental pollution but also helps in sustainable agricultural processes (Zhang et al. 2021; Elbanna et al. 2024).

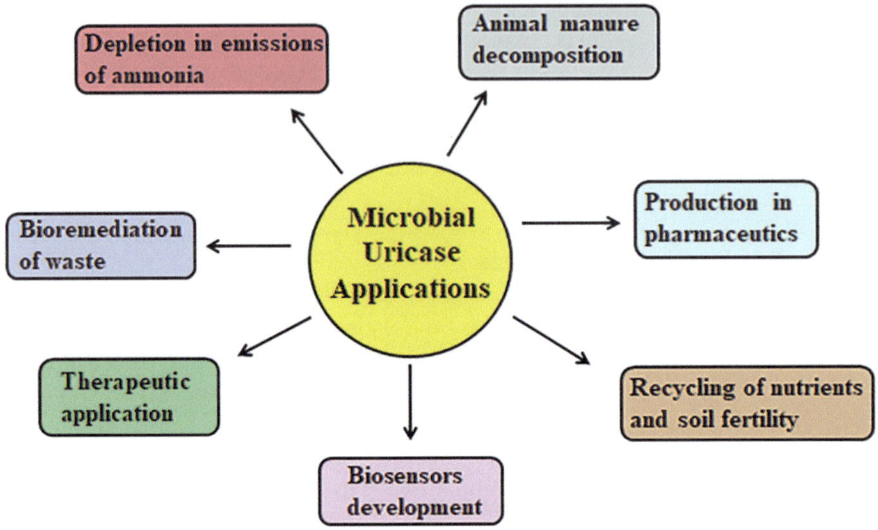

Fig. 3.10 Illustration of various applications of microbial uricase

Reduction of Ammonia Emissions The uricase activity is required to minimize ammonia emissions released when manure is decomposing and suppresses uric acid buildup. The relevance is that ammonia emissions may lead to air pollution, which in turn promotes acid rain development (Elbanna et al. 2024).

3.7.2 Industrial Applications

Pharmaceutical, textile, and food industries, which generate nitrogenous waste, usually experience difficulties in the management of uric acid accumulation. Microbial uricase is a green method of uric acid biodegradation in industrial wastewater, thereby removing toxicity and environmental concerns (Gupta et al. 2022). Microbial uricase is employed in biosensors that measure the concentration of uric acid in environmental and clinical samples. The biosensors play an important role in the evaluation of kidney function, diagnosis of metabolic disorders, and evaluation of water quality in contaminated environments (Ghosh et al. 2015; Chen et al. 2023). Uricase is also significant in the pharmaceutical industry, especially in the management of hyperuricemia and gout. Mass production of microbial uricase from microorganisms such as *Aspergillus niger* and *Pseudomonas aeruginosa* has enabled the production of uricase-based drugs, for instance, rasburicase for the treatment of high uric acid in chemotherapy (Wang et al. 2021). Microbial uricase, while promising in most uses, still faces numerous challenges in terms of enzyme stability, production cost, and regulatory approval for commercial production on a large scale (Hassan et al. 2023).

3.7.3 Therapeutic Applications of Uricase

Evolutionary adaptations have led to the uricase gene inactivation in humans, resulting in hyperuricemia above a critical level. When the concentration of uric acid is above this critical level, it forms sodium urate crystals in soft tissues and causes gout. The disease is expressed in the form of acute flare-ups, tophi, and urate arthropathy. Gout is defined as the most prevalent inflammatory arthritis and affects 1%–2% of the population in industrialized nations (Briere et al. 2002; Richette et al. 2007; Patte et al. 2002). Therapeutically, microbial uricase, an enzyme that catalyzes the oxidation of uric acid to allantoin, has been of immense clinical significance. Its versatility of therapeutic usefulness speaks to its use in modern medicine, and the following therapeutic applications.

(i) *Uricase Implications in Biosensors: Advanced Approaches for Uric Acid Detection*

Detection of uric acid in urine and blood is crucial in the diagnosis of numerous health ailments. Several methods are employed to estimate uric acid, including mass fragmentography, enzyme electrodes, colorimetry, radiochemical-HPLC, fluorescent sol-gel techniques, uric acid commercial kits, and chemiluminescence. Of these, the colorimetric method has been famous for its simplicity, specificity, and sensitivity since it employs peroxidase as well as uricase. The enzymes are, however, expensive, and the method is therefore expensive compared to a few others. To address this, innovative biosensing techniques have been developed. These approaches involve immobilizing uricase on electrode surfaces using advanced materials like ZnO nanorods, polyaniline-polypyrrole films, ZnO nano-flakes, and polypyrrole nano-electrodes. Such biosensors offer an inexpensive and efficient means of monitoring the concentration of uric acid, paving the way to improved disease diagnosis and personalized medicine.

A notable development involves the creation of a colorimetric sensor that integrates uricase with a chromium-doped vanadium dioxide (Cr/VO_2) nanozyme and illustrated in Fig. 3.11. The hybrid system takes advantage of the peroxidase-mimicking activity of Cr/VO_2 to catalyze reactions effectively, allowing for the detection of uric acid with a low detection limit of 0.34 μM. The sensor operates effectively at room temperature, making the testing process easy and promising for clinical use (Kuang et al. 2025).

Another novel strategy entails incorporating uricase within copper nanozymes. The copper-based nanozymes resemble the catalytic behavior of horseradish peroxidase and have proven effective for uric acid-detection. By adding uricase, these biosensors exhibit enhanced selectivity and sensitivity while serving as an economical and effective means for tracking uric acid concentrations (Chi et al. 2022). Furthermore, a report was made on the development of an ultrasensitive biosensor for uric acid utilizing a second-generation design. This biosensor boasts a fast response time of approximately 2 seconds and has been successfully used for the analysis of human blood samples, indicating its potential for use in point-of-care

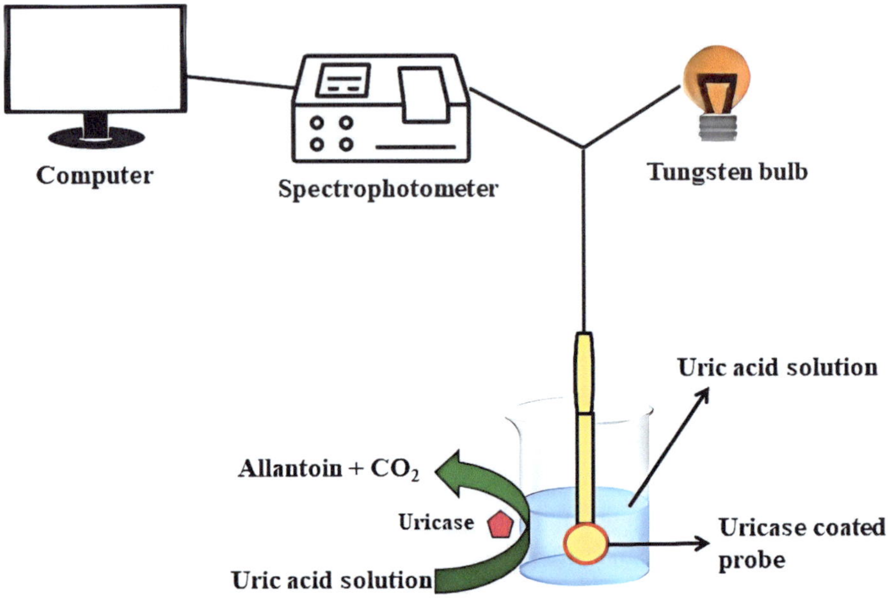

Fig. 3.11 Illustration of the experimental setup for uric acid detection using a micro-ball fiber structure

testing applications (Deffo et al. 2023). These advancements highlight the significant potential of uricase-based biosensors to provide fast, sensitive, and cost-effective uric acid detection, hence enhancing disease diagnosis and personalized medicine.

(ii) *Application for the Treatment of Hyperuricemia and Gout*

Hyperuricemia, which refers to serum urate levels >6.8 mg/dL, is the major cause of gout, an arthritis associated with monosodium urate crystal deposition within joints and soft tissues, which results in profound pain and inflammation (Bardin and Richette 2017). Current management options consist of oral urate-lowering therapies (ULTs), which treat to decrease serum uric acid and prevent crystallization (Dalbeth et al. 2019). Uricase has become a new therapeutic enzyme for treating different forms of gout (Fig. 3.12). Pegloticase, a PEGylated uricase, is approved for the treatment of chronic refractory gout. It has shown effectiveness in significantly reducing serum uric acid (SUA) levels and resolving tophi (Sherman et al. 2008).

PEGylation improves uricase's effectiveness in treating chronic gout by attaching polyethylene glycol (PEG) molecules to it. This modification increases the enzyme's stability, prolongs its action, and reduces immune system reactions (Bomalaski and Clark 2004). However, some patients develop immune responses against PEG, which can reduce treatment effectiveness and cause side effects (Chiu

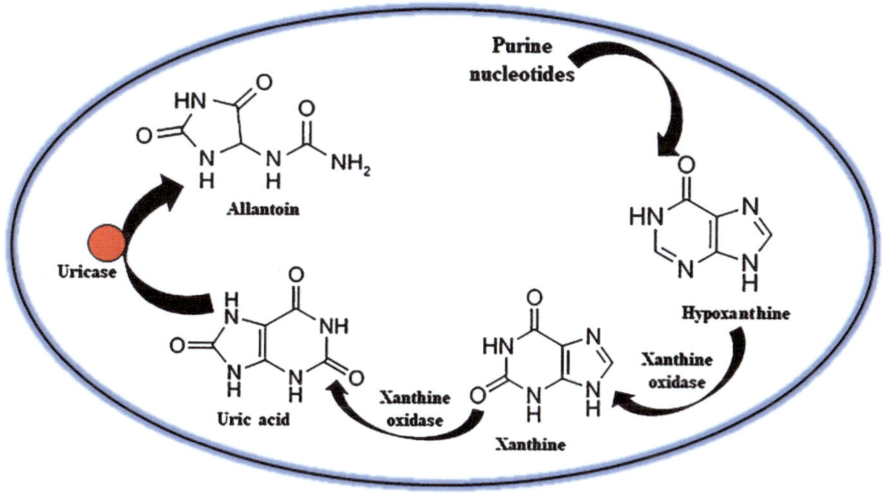

Fig. 3.12 Diagrammatic illustration of the application of uricase enzymatic treatment for hyperuricemia and gout

2021). To address this, researchers are developing methods like site-specific PEGylation, which targets certain areas of the enzyme to reduce immune reactions (Sherman et al. 2008), and other strategies to improve the enzyme's stability and safety (Chiu 2021). These advances aim to make uricase therapies more effective and safer for patients.

(iii) *Nanozyme-Based Therapies: A New Paradigm*

Although PEGylation is a huge step forward, the constraints in terms of immunogenicity and enzymatic by-products require the development of alternative approaches. Nanotechnological breakthroughs have given rise to the creation of nanozyme-based therapies, which replicate natural enzyme activity with increased stability and catalytic performance (Huang et al. 2019). Some recent research has investigated nanozymes coupling uricase-like activity to break down uric acid with catalase-like activity to reduce the resulting H_2O_2 through a self-cascade process (Zhang et al. 2022). This bimodal function not only improves the breakdown of uric acid but also counteracts oxidative stress linked with H_2O_2 accumulation (Ye et al. 2025). Despite its efficacy, challenges such as the production of hydrogen peroxide (H_2O_2) as a by-product and potential immunogenicity have been noted (Pui et al. 2011) that require alternative therapy (Richette and Bardin 2010).

(iv) *Two-Dimensional Pd@Ir Nanosheets: A Breakthrough in Nanozyme Design*

Notably, recent research has unveiled two-dimensional Pd@Ir nanosheets as promising nanozymes that show both uricase and catalase activities (Liu et al. 2023b)

(2023). The nanosheets have been found to catalyze the decomposition of uric acid and eliminate H_2O_2 simultaneously, thus mitigating oxidative stress and improving biocompatibility (Chen et al. 2023). This is a breakthrough in the development of therapeutics using enzymes for the treatment of hyperuricemia and gout and presents a more targeted and efficient approach (Zhang et al. 2022).

(v) *Tumor Lysis Syndrome (TLS)*

Tumor lysis syndrome (TLS) is an acute and life-threatening metabolic syndrome that occurs most frequently as a complication of cancer therapy, especially in patients with hematologic malignancies like leukemia, lymphoma, and myeloma. Although TLS is possible following chemotherapy, radiotherapy, or immunotherapy, chemotherapy most commonly initiates TLS by causing abrupt and widespread lysis of tumor cells (Cairo and Bishop 2004). This is followed by a quick killing of tumor cells, which causes intracellular contents such as nucleic acids, potassium, phosphate, and uric acid to be released into the circulation. Consequently, TLS frequently results in severe metabolic derangements that involve hyperuricemia, hyperkalemia, hyperphosphatemia, and eventually renal failure, all of which contribute to increased risk of morbidity and mortality even in patients with potentially curable malignancies (Zhao et al. 2021).

TLS is diagnosed clinically if there are at least two of the following laboratory abnormalities: hyperuricemia, hyperkalemia, hyperphosphatemia, and secondary hypocalcemia (Cairo and Bishop 2004). Among them, hyperuricemia (hyperuricemia or increased uric acid level) is one of the earliest and most significant features because the excess uric acid will crystallize in the renal tubules and cause uric acid nephropathy and acute kidney injury, both of which will deteriorate the general prognosis (Cairo and Bishop 2004). The metabolic effects of TLS can lead to serious complications like arrhythmias, seizures, and multi-organ failure if not treated immediately (Jiang et al. 2022). Therefore, prevention and treatment of TLS need immediate intervention to correct these imbalances, especially hyperuricemia.

One of the best methods for treating TLS, especially in its management of the hyperuricemia aspect, is through the administration of uricase (Fig. 3.13), which catalyzes the conversion of uric acid to allantoin, a very soluble molecule that can readily be excreted by the kidney (Miele et al. 2022). Uricase readily inhibits uric acid crystallization in the kidneys, the underlying cause of uric acid nephropathy in TLS. Significantly, rasburicase, the recombinant version of uricase, has seen extensive clinical practice usage in the prevention of a decreased level of elevated uric acid in at-risk TLS patients. The enzyme's ability to lower uric acid levels quickly in hours has been a game-saver relative to conventional drugs like allopurinol, which requires a longer time to achieve therapeutic concentration and is perhaps less effective at avoiding kidney damage (Ponce et al. 2008).

Current research points to the effectiveness of rasburicase in individuals with elevated tumor burden or pre-existing renal dysfunction, both of which are independent predictors of severe TLS (Strader et al. 2004). One of the major benefits of rasburicase is its quick onset of action, lowering uric acid levels within minutes,

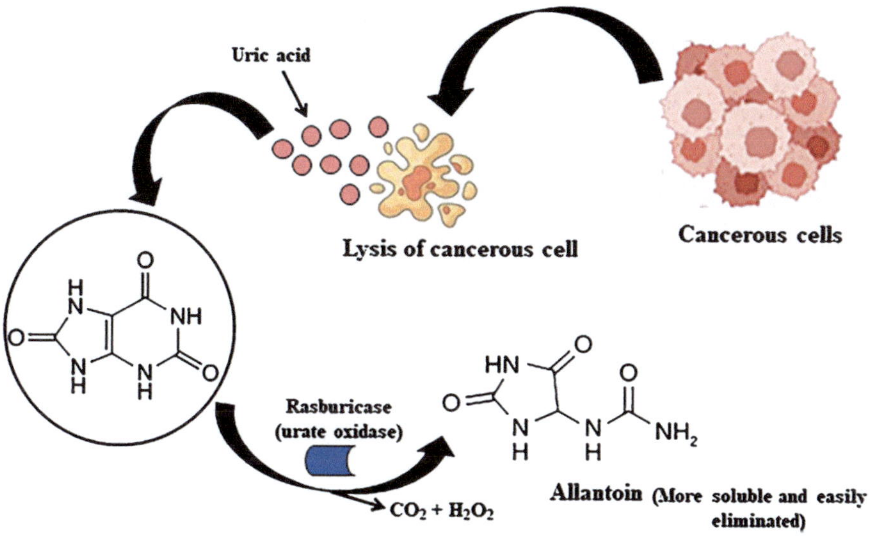

Fig. 3.13 Illustration depicting the therapeutic action of uricase against TLS

which is essential in avoiding renal injury and other life-threatening complications of TLS (Coiffier et al. 2008). For example, rasburicase has been demonstrated to decrease the occurrence of acute kidney injury and dialysis requirements in TLS patients (Abdel-Nabey et al. 2022), and it is a critical tool in avoiding TLS complications in high-risk groups.

3.8 Conclusions

Microbial uricase from sources such as *Aspergillus niger*, *Pseudomonas aeruginosa*, *Bacillus subtilis*, and *Aspergillus flavus* has various applications in agriculture, industry, and medicine. Microbial uricase oxidizes uric acid to nitrogenous compounds, increasing soil fertility and facilitating sustainable agriculture. In foods and beverages, it is responsible for quality control, whereas in cosmetics and drugs, it assists in the production of allantoin. Clinically, uricase has revolutionized the treatment of hyperuricemia and gout, with PEGylated versions such as pegloticase providing better stability and fewer immune responses. In cancer, rasburicase is used to avert kidney injury by lowering uric acid in tumor lysis syndrome (TLS) rapidly. Developments in uricase-based nanozymes are enhancing catalytic activity and biocompatibility, holding the prospect of next-generation treatments. Despite challenges in stability, affordability, and regulation, continued research in nanotechnology and genetic engineering continues to grow its scope in medicine, environmental stewardship, and industry.

References

Abdel-Fattah YR, Hafez EE, El-Helow ER (2005) Uricase production by *Penicillium* species: characterization and industrial potential. J Ind Microbiol Biotechnol 32(1):1–7. https://doi.org/10.1007/s10295-004-0193-7

Abdel-Nabey M, Chaba A, Serre J, Lengline E, Azoulay E, Darmon M, Zafrani L (2022) Tumor lysis syndrome, acute kidney injury, and disease-free survival in critically ill patients requiring urgent chemotherapy. Ann Intensive Care 12:15. https://doi.org/10.1186/s13613-022-00990-1

Aly M, Tork S, Al-Garni S, Allam R (2013) Production and characterization of uricase from *Streptomyces exfoliatus* UR10 isolated from farm wastes. Turk J Biol 37(5):3. https://doi.org/10.3906/biy-1206-3

Ames BN, Cathcart R, Schwiers E, Hochstein P (1981) Uric acid provides an antioxidant defense in humans against oxidant- and radical-caused aging and cancer: a hypothesis. Proc Natl Acad Sci USA 78(11):6858–6862. https://doi.org/10.1073/pnas.78.11.6858. PMID: 6947260; PMCID: PMC349151.

Argyrou E, Sophianopoulou V, Schultes N, Diallinas G (2001) Functional characterization of a maize purine transporter by expression in *Aspergillus nidulans*. Plant Cell 13:953–964

Atalla SMM, El-Shamy AR (2013) Optimization of uricase by *Aspergillus terreus* NRC 7. World Appl Sci J 23(7):871–876. https://doi.org/10.5829/idosi.wasj.2013.23.07.13101

Atalla MM, Farag MM, Eman RH, Abd-El-Lataif MS, Nehad EA (2009) Optimum conditions for uricase enzyme production by *Gliomastix gueg*. Malays J Microbiol 5(1):45–50

Bakhtiari S, Toosi P, Samadi S, Bakhshi M (2017) Assessment of uric acid level in the saliva of patients with oral lichen planus. Med Princ Pract 26:57–60

Barbieri F et al (2020a) Anaerobic oxidation of formate and uric acid reduction in gut microbiota: a comprehensive review. J Microb Biochem 34(2):200–210

Barbieri F et al (2020b) Uricase and its role in uric acid degradation in fungi and animals. J Microb Biochem 34(2):200–210

Bardin T, Richette P (2017) Definition of hyperuricemia and gouty conditions. Nat Rev Rheumatol 13(2):105–106. https://doi.org/10.1038/nrrheum.2017.3

Barker HA (1943) *Streptococcus allantoicus* and the fermentation of allantoin. J Bacteriol 46(3):251–259. https://doi.org/10.1128/jb.46.3.251-259.1943

Bomalaski JS, Clark WR (2004) PEGylation: a strategy for enhancing the pharmacokinetic properties and reducing the immunogenicity of therapeutic proteins. J Clin Rheumatol 10(3):139–146

Bongaerts GP, Vogels GD (1976) Uric acid degradation by *Bacillus fastidiosus* strains. J Bacteriol 125:689–697. https://doi.org/10.1128/jb.125.2.689-697.1976

Briere C, Richette P, Hoenen-Clavert V, Loeuille D, Bardin T (2002) Urate-oxidase in the prevention and treatment of metabolic complications in patients with B-cell lymphoma and leukemia, treated in the Société Française d'Oncologie Pédiatrique LMB89 protocol. Ann Oncol 13(5):789–795

Cairo MS, Bishop M (2004) Tumour lysis syndrome: new therapeutic strategies and classification. Br J Haematol 127(1):3–11. https://doi.org/10.1111/j.1365-2141.2004.05094

Chen MH, Li SY (2023) Extra-cellular production of uricase through the sec-type secretion system in *Escherichia coli*. Biochem Eng J 194:108894

Chen C, Lu JM, Yao Q (2016) Hyperuricemia-related diseases and xanthine oxidoreductase (XOR) inhibitors: an overview. Med Sci Monit 2016(22):2501–2512

Chen S, Zhou Y, Wu J (2023) Catalytic dual-functionality of Pd@Ir nanosheets in hyperuricemia therapy. ACS Nano 17(9):8242–8253. https://doi.org/10.1021/acsnano.3c03820

Chi X, Cheng Q, Yang K et al (2022) Fabrication of a novel nano-biosensor for efficient colorimetric determination of uric acid. Appl Nanosci 12:2255–2264. https://doi.org/10.1007/s13204-022-02498-3

Chiu YT (2021) Advances in PEGylation of uricase for the treatment of gout. Gout Hyperuricemia J 15(2):43–56

Coiffier B, Altman A, Pui CH, Younes A, Cairo MS (2008) Guidelines for the management of pediatric and adult tumor lysis syndrome: an evidence-based review. J Clin Oncol 26(16):2767–2778. https://doi.org/10.1200/JCO.2007.15.0177

Dalbeth N, Merriman TR, Stamp LK (2019) Gout. Lancet 393(10183):1300–1312. https://doi.org/10.1016/S0140-6736(18)33293-8

Dawson J, Walters M (2006) Uric acid and xanthine oxidase: future therapeutic targets in the prevention of cardiovascular disease. Br J Clin Pharmacol 62(6):633–644. https://doi.org/10.1111/j.1365-2125.2006.02729

Deffo G, Hazarika R, Cedric M, Ngaha D, Basumatary M, Kalita S, Puzari P, Ngameni E (2023) An ultra-sensitive uric acid second generation biosensor based on chemical immobilization of uricase on functionalized multiwall carbon nanotube grafted palm oil fiber in the presence of a ferrocene mediator. Anal Methods 28(20):2401–2538

Desai A, Kumar R, Sharma P (2023) Purine metabolism and its disorders: a molecular perspective. Metab Pathways Rev 45(1):78–92. https://doi.org/10.1016/j.mpr.2023.07.011

Desideri G, Castaldo G, Lombardi A, Mussap M, Testa A, Pontremoli R, Ferri C (2014) Is it time to revise the normal range of serum uric acid levels? Eur Rev Med Pharmacol Sci 18:1295–1306

Elbanna K, Alshareef A, Leena A, Neyaz M, Mahmoud Z, El-Readi HH, Abulreesh H (2024) Microbial uricase and its unique potential applications. Adv Microbiol 63(2):81–91. https://doi.org/10.2478/am-2024-0007

French JB, Ealick SE (2011) Structural and kinetic insights into the mechanism of 5-hydroxyisourate hydrolase from *Klebsiella pneumoniae*. Acta Crystallogr D Biol Crystallogr 67(Pt 8):671–677. https://doi.org/10.1107/S090744491101746X

Geweely NS, Nawar LS (2011) Production, optimization, purification, and properties of uricase isolated from some fungal flora in Saudi Arabian soil. Aust J Basic Appl Sci 5:220–230

Ghosh T, Sarkar P, Anthony PF, Turner (2015) A novel third-generation uric acid biosensor using uricase electro-activated with ferrocene on a Nafion-coated glassy carbon electrode. Bioelectrochemistry 102:1–9

Gupta R, Sharma S, Kumar P (2022) Industrial applications of uricase: current trends and future prospects. Biotechnol Adv 41:107689

Hassan M, Ali Z, Rahman A (2023) Improving microbial uricase stability and production using genetic engineering. Appl Microbiol Biotechnol 107(3):679–692

Huang Y, Ren J, Qu X (2019) Nanozymes: classification, catalytic mechanisms, activity regulation, and applications. Chem Rev 119:4357–4412. https://doi.org/10.1021/acs.chemrev.8b00672

Jiang X et al (2022) Uric acid metabolism in bacteria and plants: enzymatic pathways and functional implications. Environ Microbiol 24(5):1135–1146

Kahn PJ et al (2021) The biochemical and genetic basis of uric acid metabolism in bacteria and fungi. Ann Rev Microbiol 75:125–143

Kanbay M, Segal M, Afsar B, Kang DH, Rodriguez-Iturbe B, Johnson RJ (2013) The role of uric acid in the pathogenesis of human cardiovascular disease. Heart 99:759–766

Kim YS et al (2000) Regulation of purine metabolism genes in *E. coli*: role of PurR and the hpx operon. J Bacteriol 182(9):2533–2542

Kim HS et al (2004) Characterization of uric acid metabolism in *Klebsiella pneumoniae*. Microb Pathog 37(3):125–131

Kim NY, Lee YJ, Park JW, Kim SN, Kim EY, Kim Y, Kim OB (2021) An Escherichia coli FdrA variant derived from syntrophic coculture with a methanogen increases succinate production due to changes in allantoin degradation. mSphere 6. https://doi.org/10.1128/msphere.00654-21

Kratzer FJT et al (2014) Evolutionary history and metabolic insights of ancient mammalian uricases. Proc Natl Acad Sci USA 111:3763–3768

Kuang G, Ke W, Gu C et al (2025) Enzyme cascade nanozyme-based colorimetric sensor for detection of uric acid as a biomarker of hyperuricemia. Microchim Acta 192:85. https://doi.org/10.1007/s00604-024-06946-5

Kuwabara M, Fukuuchi T, Aoki Y, Mizufa E, Ouchi M, Kurajoh M, Maruhashi T, Tanaka A, Morikawa N, Nishimiya K, Akashi N, Tanaka Y, Oani N, Morita M, Miyata H, Takada T, Tsutani H, Ogino K, Ichida K, Hisatome I et al (2023) Exploring the multifaceted nexus of uric acid and health: a review of recent studies on diverse diseases. Biomolecules 13:1519

Lee IR, Yang L, Sebetso G, Allen HN, R., Doan Thi, Ross B et al (2013) Characterization of the complete uric acid degradation pathway in the fungal pathogen *Cryptococcus neoformans*. PLoS One 8(5):1–13

Li W, Xu S, Zhang B, Zhu Y, Hua Y, Kong X et al (2017) Directed evolution to improve the catalytic efficiency of urate oxidase from *Bacillus subtilis*. PLoS One 12(5):e0177877. https://doi.org/10.1371/journal.pone.0177877

Li Z et al (2021) Urate oxidase and its enzymatic network in archaea and bacteria. Front Cell Infect Microbiol 11:345–356

Li D, Zhang M, La ALTZ, Lyu Z, Li X, Feng Y, Liu D, Guo Y, Hu Y (2023) Quercetin-enriched *lactobacillus aviaries* alleviated hyperuricemia by hydrolase-mediated degradation of purine nucleosides. Pharmacol Res 196:106928

Liu Y, Jarman JB, Low YS, Augustijn HE, Huang S, Chen H, DeFeo ME, Sekiba K, Hou BH, Meng X, Weakley AM, Cabrera AV, Zhou Z, van Wezel G, Medema MH, Ganesan C, Pao AC, Gombar S, Dodd DA (2023a) A widely distributed gene cluster compensates for uricase loss in hominids. Cell 186(16):3400–3413.e20., ISSN 0092-8674. https://doi.org/10.1016/j.cell.2023.06.010

Liu Y, Li Z, Wang X (2023b) Two-dimensional Pd@Ir nanosheets as bifunctional nanozymes for uric acid degradation and hydrogen peroxide elimination. Adv Funct Mater 33(14):2300134. https://doi.org/10.1002/adfm.202300134

Liu W, Nan F, Liu F, Yang X, Li Z, Jiang S, Zhang X, Li J, Yu M, Wang Y, Wang B (2024) Isolation and identification of uric acid-dependent *Aciduricibacillus chroicocephali* gen. nov., sp. nov. from seagull feces and implications for hyperuricemia treatment. mSphere 9:e 00025-24. https://doi.org/10.1128/msphere.00025-24

Loffler W, Grobner W, Medina R, Zollner N (1982) Influence of dietary purines on pool size, turnover, and excretion of uric acid during balance conditions. Isotope studies using 15N-uric acid. Res Exp Med 181:113–123. https://doi.org/10.1007/BF01852188

Lotfy WA (2008) Production of a thermostable uricase by a novel *Bacillus thermocatenulatus* strain. Med Princ Pract 99(699):702

Lu N, Dubreuil M, Zhang Y, Neogi T, Rai SK, Ascherio A, Hernan MA, Choi HK (2016) Gout and the risk of Alzheimer's disease: a population-based, Bmi-matched cohort study. Ann Rheum Dis 75:547–551. https://doi.org/10.1136/annrheumdis-2014-206917

Maiuolo J, Oppedisano F, Gratteri S, Muscoli C, Mollace V (2016) Regulation of uric acid metabolism and excretion. Int J Cardiol 213:8–14. https://doi.org/10.1016/j.ijcard.2015.08.109

Mandal T, Basu A, Choudhury S (2022) The dual role of uric acid in human health: biochemical significance and clinical implications. J Biochem Res 38(2):115–123. https://doi.org/10.1016/j.jbr.2022.05.004

Manimekalai V, Anandharaj B (2021) Optimization of growth parameters for Uricase enzyme activity of *Bacillus subtilis* MM13 enumerated from birds fecal contaminated soil. Bull Environ Pharmacol Life Sci 9:77–84

Matsumoto H, Ohta S, Kobayashi R, Terawaki Y (1978) Chromosomal location of genes participating in the degradation of purines in *Pseudomonas aeruginosa*. Mol Gen Genet 167(1978):165–176

Meletis CD, Barker JE (2005) Therapeutic uses of amino acids. Altern Ther Health Med 11:24–28

Miele E, Dose WM, Manyakin I, Frosz M, Ruff Z, De Volder MFL et al (2022) Hollow-core optical fibre sensors for operando Raman spectroscopy investigation of Li-ion battery liquid electrolytes. Nat Commun 13:1651. https://doi.org/10.1038/s41467-022-29330-4

Moreno D et al (2023a) Formate dehydrogenase in uric acid degradation pathways: mechanistic insights. Microb Ecol 66(5):1224–1235

Moreno D et al (2023b) Comprehensive review of uric acid degradation in bacteria and fungi. Microb Ecol 66(5):1224–1235

Negm E-DA, Ezzat A, Aly HF, Younis EA, Awad GA, Farid MAM (2023) Hypouricemic, anti-inflammatory, and antioxidant activities of *lactobacillus*-based functional yogurt in induced-arthritic male Wistar rats: therapeutic and protective potentials. Biocatal Agric Biotechnol 47:102597

Nguyen CX, Dohnalkova A, Hancock CN, Kirk KR, Stacey G, Stacey MG (2023) Critical role for uricase and xanthine dehydrogenase in soybean nitrogen fixation and nodule development. Plant Genome 16(2):e20171. https://doi.org/10.1002/tpg2.20172

Ogawa J (2006) Analysis of microbial purine metabolism and its application for hyperuricemia prevention. Institute for Scientific Research Grant, Noda

Parvez S et al (2016) *Pseudomonas spp*. in bioremediation: uric acid degradation potential. Environ Microbiol 18(1):40–49

Patte C, Sakiroglu C, Ansoborlo S, Baruchel A, Plouvier E, Pacquement H, Babin-Boilletot A (2002) Urate-oxidase in the prevention and treatment of metabolic complications in patients with B-cell lymphoma and leukemia, treated in the Societe Française d'Oncologie Pediatrique LMB89 protocol. Ann Oncol 13(5):789–795

Perez-Ruiz F, Martínez-Indart L, Carmona L, Herrero-Beites AM, Pijoan JI, Krishnan E (2014) Tophaceous gout and high level of hyperuricaemia are both associated with an increased risk of mortality in patients with gout. Ann Rheum Dis 73:177–182

Ponce D, Liu D, Rasul M, Ahmed N, Dimicco M, Seiter K (2008) Rasburicase effectively lowers serum uric acid levels in high-risk, poor-performance status patients with hematologic malignancies. Blood 112(11):4012. https://doi.org/10.1182/blood.V112.11.4012.4012

Pui CH, Jeha S, Cheng C (2011) Recombinant uricase in the management of hyperuricemia in leukemia. Blood 117(6):1675–1682. https://doi.org/10.1182/blood-2010-09-310987

Pustake SO, Bhagwat PK, Dandge PB (2019) Statistical media optimization for the production of clinical uricase from *Bacillus subtilis* strain SP6. Heliyon 5(5):e01756. https://doi.org/10.1016/j.heliyon.2019.e01756

Rajagopalan P, Chandramoorthy HC, Elbessoumy AA (2017) Pilot scale production and characterization of uricase from Penicillium purpurescens. Ind J Biotechnol 16:570–577. http://nopr.niscpr.res.in/handle/123456789/44374

Ramazzina I, Folli C, Secchi A, Berni R, Percudani R (2006) Completing the uric acid degradation pathway through phylogenetic comparison of whole genomes. Nat Chem Biol 2(3):144–148. https://doi.org/10.1038/nchembio768

Ravichandran R, Hemaasri S, Swaranjit S, Cameotra SS, Jayaprakash NS (2015) Purification and characterization of an extracellular uricase from a new isolate of *Sphingobacterium thalpophilum* (VITPCB5). Protein Expr Purif 114:136–142, 1046–5928. https://doi.org/10.1016/j.pep.2015.06.017

Richette P, Bardin T (2010) Gout. Lancet 375(9711):318–328. https://doi.org/10.1016/S0140-6736(09)60883-7

Richette P, Briere C, Hoenen-Clavert V, Loeuille D, Bardin T (2007) Rasburicase for tophaceous gout not treatable with allopurinol: an exploratory study. J Rheumatol 34(10):2093–2098. PMID: 17896799

Ridi ER, Tallima H, Migliardo F (2017) Biochemical and biophysical methodologies open the road for effective schistosomiasis therapy and vaccination. Biochim Biophys Acta 1861:3613–3620

Roman YM (2023) The role of uric acid in human health: insights from the Uricase Gene. J Pers Med 13:1409. https://doi.org/10.3390/jpm13091409

Saeed HM, Yousry YRAF, Gohar M, Elbaz MA (2004) Purification and characterization of extracellular *Pseudomonas aeruginosa* urate oxidase enzyme. Pol J Microbiol 53:045–052

Sherman DM et al (2008) Site-specific PEGylation of uricase: a strategy to reduce immunogenicity. J Biotechnol 133(1):36–44

Smith D et al (2021) Evolutionary insights into uricase and its absence in humans. Nat Rev Gastroenterol 19(7):498–510

Strader D, Oholendt MS, Popat U, Carrum G, Heslop HE (2004) Rasburicase improves renal function and uric acid nephropathy in patients developing acute renal failure after conditioning for hematopoietic stem cell transplant (HSCT). Blood 104(11):1142. https://doi.org/10.1182/blood.V104.11.1142.1142

Suh MJ et al (2005) Metabolic pathways of uric acid degradation by *Candida utilis*. FEMS Yeast Res 6(2):256–263

Sumner JB, Somers GF (1947) Chemistry and methods of enzymes. Academic Press, Inc., New York. Varner, J. E. *1959*. *Urease*, 4, 247–256.

Suzuki K, Sakasegawa SI, Misaki H, Sugiyama M, Ravichandran (2004) Molecular cloning and expression of uricase gene from *Arthrobacter globiformis* in *Escherichia coli* and characterization of the gene product. J Biosci Bioeng 98:153–158

Tandon S, Sharma A, Singh S, Sharma S, Sarma SJ (2021) Therapeutic enzymes: discoveries, production and applications. J Drug Deliv Sci Technol 63:102455. https://doi.org/10.1016/j.jddst.2021.102455

Valentine RC, Bojanowski R, Elizabeth G., Wolfe RS (1962) Mechanism of the Allantoin fermentation. Eur J Biochem 237(7):2271–2277

Varner JE (1959) Urease. In: Boyer PD, Lardner H, Myrback K (eds) The enzymes, vol 4. Academic Press, Inc., New York, pp 247–256

Vidal MJ et al (2022a) Comparative genomics of purine metabolism in gram-negative bacteria. Mol Biol Rep 49(5):412–426

Vidal M et al (2022b) Genetic and enzymatic regulation of uric acid metabolism in anaerobic environments. J Bacterial Biochem 11(6):455–467

Vogels GD, Van der Drift C (1976) Degradation of purines and pyrimidines by microorganisms. Bacteriol Rev 40(2):403–468. Retrieved from https://pubmed.ncbi.nlm.nih.gov/776950.

Wan S, Liu X, Sun W, Lv B, Li C (2023) Current advances for omics-guided process optimization of microbial manufacturing. Bioresour Bioprocess 10:30

Wang S et al (2021) Role of xanthine oxidoreductase and urate reductase in formate-dependent degradation of uric acid. J Microbiol Metab 22(3):177–188

Wang Y, Zhang X, Chen L, Li J (2024) Microbial uricase and its unique potential applications. Acta Microbiol Immunol Hung 71(1):85–95

Watanabe S, Kang D-H, Feng L, Nakagawa T, Kanellis J, Lan H, Mazzali M, Johnson RJ (2002) Uric acid, hominoid evolution, and the pathogenesis of salt-sensitivity. Hypertension 40:355–360

Wei G et al (2022a) Formate-dependent uric acid degradation in human gut microbiota: the key metabolic processes and pathways. Environ Microbiol 24(9):4401–4410

Wei W et al (2022b) Nitrogen utilization and metabolic adaptation in anaerobic fungi. FEMS Microbiol Ecol 98(7):090

Wei G et al (2022c) Ureidoglycolate hydrolase and nitrogen recycling in uric acid metabolism. Environ Microbiol 24(9):4401–4410

Wen S, Arakawa H, Tamai I (2023) Uric acid in health and disease: from physiological functions to pathogenic mechanisms. Pharmacol Ther 256:108615. https://doi.org/10.1016/j.pharmthera.2024.108615

Wu XW, Lee CC, Muzny DM, Caskey CT (1989) Urate oxidase: primary structure and evolutionary implications. Proc Natl Acad Sci USA 86(23):9412–9416

Wu X, Muzny DM, Lee CC, Caskey CT (1992) Two independent mutational events in the loss of urate oxidase during hominoid evolution. J Mol Evol 34:78–84

Wu X, Wakamiya M, Vaishnav S, Geske R, Montgomery C, Jones P, Caskey CT (1994) Hyperuricemia and urate nephropathy in urate oxidase-deficient mice. Proc Natl Acad Sci USA 91(2):742–746. https://doi.org/10.1073/pnas.91.2.742

Xu X, Yan Y, Huang J, Zhang Z, Wang Z, Wu M, Liang H (2022) Regulation of uric acid glyoxylate metabolism by Ugm R protein in *Pseudomonas aeruginosa*. Environ Microbiol 24:3242–3255

Ye Z, Wang Y, Zhang G, Hu X, Wang J, Chen X (2025) Exploration of uricase-like activity in Pd@Ir nanosheets and their application in relieving acute gout using self-cascade reaction. J Colloid Interface Sci 678:380–392. https://doi.org/10.1016/j.jcis.2024.09.140

Yeldandi AV, Vijay Y, Sujata K, Narasimha MCV, Xuedong W, Alvares K et al (1991) Molecular evolution of the urate oxidase-encoding gene in hominoid primates: nonsense mutations. Gene 109(2):281–284

Yoshimoto T et al (1991) Intracellular uricase activity in *Micrococcus luteus*. Appl Microbiol Biotechnol 35(1):100–105

Zhang T, Zhao L, Chen W (2021) Microbial decomposition of uric acid in poultry waste: enzymatic and microbial approaches. Environ Microbiol Rep 13(2):157–169

Zhang L, Chen Y, Li H (2022) Advances in nanozyme-based therapies for gout: dual-functional uricase and catalase mimetics. Nano Res 15(7):5771–5782. https://doi.org/10.1007/s12274-022-4231-8

Zhao H, Wu L, Yan G, Chen Y, Zhou M, Wu Y, Li Y (2021) Inflammation and tumor progression: signaling pathways and targeted intervention. Signal Transduct Target Ther 6(1):263. https://doi.org/10.1038/s41392-021-00658-5

Zhou XL, Ma XH, Sun GQ, Li X, Guo KP (2005) Isolation of a thermostable uricase-producing bacterium and study on its enzyme production conditions. Process Biochem 40:3749–3753

Streptokinase Evolution and Its Pharmaceutical Applications

4

Richa Salwan, Rhydum Sharma, Randhir Kaur,
Vivek Sharma, and Mansour Ghorbanpour

Abstract

The formation of clots, also known as "thrombosis," restricts the supply of blood that can cause acute myocardial infarction and ischemic stroke. The reduced blood flow can cause death. The therapeutic measures involving the surgical interventions, such as removing or bypass the blockage, developing alternate collateral vessels for blood supply, and deploying the thrombolytic enzymes such as streptokinase, are an attractive choice. The pharmaceutical applications of streptokinase include treating acute myocardial infarction, acute retinal vein occlusion, arterial thrombosis and embolism, and deep vein thrombosis. Streptokinase is also used fo the revacularization and restoration of antegrade or forward blood flow to enahnce the tissue survival and patient health. Being an indirect fibrinolytic agent, streptokinase interacts with plasminogen to form an active complex with proteolytic activity that can turn plasminogen into plasmin. The activated complex can resolve the blood clots that obstruct the coronary arteries. The early usage of streptokinase formulations was reported for restricting thrombolytic potency. However, adverse effects such as immunogenicity comparatively short half-life, tissue damage, and peripheral bleeding were major concerns. Due to immense potential of streptokinase, researchers continuously

R. Salwan
College of Horticulture and Forestry, Dr YS Parmar University of Horticulture & Forestry, Hamirpur, Himachal Pradesh, India

R. Sharma · R. Kaur · V. Sharma (✉)
University Center for Research and Development, Chandigarh University, Mohali, Punjab, India

M. Ghorbanpour (✉)
Department of Medicinal Plants, Faculty of Agriculture and Natural Resources, Arak University, Arak, Iran
e-mail: m-ghorbanpour@araku.ac.ir

© The Author(s), under exclusive license to Springer Nature Singapore Pte Ltd. 2026

V. Sharma, R. Salwan (eds.), *Pharmaceutically Relevant Microbial Enzymes*,
https://doi.org/10.1007/978-981-95-4941-2_4

worked on the enhancing the quality of the formulation using this enzyme. This chapter highlights the mechanism, applications, and challenge associated with streptokinase applications in pharmaceutical sectors.

Keywords

Blood clot · Protease activity · Human health · Benefits

4.1 Introduction

The formation of blood clot in the blood vessels is one of the leading factors of life-threatening diseases such as cardiovascular diseases. These clots are formed the formation of isopeptide bonds in αA-, βB-, and ϒ-peptide chains of fibrinogen and blood coagulation factor XIIIa. These clots are targeted for breakdown using different anticoagulant agents, antiplatelet drugs, fibrinolytic enzymes, and surgical operations. Several many studies have been done on targeting blood clots for the treatment of thromboembolysis (Shen et al. 2021). Hemorrhoids (Hernández-Bernal et al. 2015), myocardial infarction (Ojha and Dhamoon 2023), and other diseases are associated with blood clotting. The term "thromboembolism" is a circulating blood clot. Thromboembolic diseases in the form of deep vein thrombosis and pulmonary embolism can obstruct the flow of blood in veins (Khan et al. 2021). The hemorrhoids usually represent painless area with tendency to bleed and contain swollen veins in terminal part of rectum. At a global scale, thrombo-occlusion is one of the major reasons associated with morbidity and mortality. Therefore, the restoration of normal blood flow is needed for tissue survival and ultimately patient health (Holt and Gupta 2012). The fibrinolytic enzymes from microorganisms including bacteria, fungi, and others such as streptokinase have been tried to address the issue. In this context, streptokinase from *Streptococcus hemolyticus,* nattokinase from *Bacillus subtilis* YF 38, bafibrinase from *Bacillus* sp. AS-S20-I, longolytin from *Arthrobotrys longa*, versiase from *Aspergillus versicolor* ZLH-1 have been explored (Singh et al. 2023). Corticosteroids such as hydrocortisone acetate are used for the treating acute hemorrhoids (Gupta 2007). The clinical studies have shown that streptokinase suppositories resulted in significant benefits compared to hydrocortisone acetate preparation against acute hemorrhoidal illness (Hernández-Bernal et al. 2015). For example, the rectal route suppositories of 200,000 international units (IU) of streptokinase at different periodic intervals not only led to relief from pain and bleeding but is also cause more than 70% reduction of the lesion size (Hernández-Bernal et al. 2015).

Streptokinase lacks the criteria commonly required for the categorization of a protein as an enzyme (Reddy 1988). Streptokinase, an extracellular protein, has been isolated from various hemolytic streptococci strain. Initially, streptokinase has been explored for dissolving the thrombi under experimental and clinical conditions (Sherry 1969; Hirsh 1970). Although, streptokinase was introduced in the first half of the nineteenth century as a thrombolytic agent, initially it remained a failure. It gained the attention and success in the 1980s in early clinical studies for treating acute myocardial infarction (Sherry 1969; Reddy 1988). It has been observed that the early administration of streptokinase (within the first 6 h) to acute myocardial infarction patients reduced the rate of mortality. The effectiveness of streptokinase in dissolving thrombi and restoring blood flow to ischemic myocardium was attributed to its ability to activate plasminogen, responsible for enhancing fibrinolysis (Tisdale et al. 1989). Streptokinase is a 47-kilodalton extracellular protein produced by beta-hemolytic streptococci. The activation of plasminogen is achieved after the proteolytic cleavage of the bond between Arg561 and Val562. Thereafter the formation of N-terminal amino group of Val (at 562) constitutes a salt bridge with Asp (at 740 position) and creates active plasmin with protease activity (Wang et al. 1999). The active plasmin acts as a non-specific protease and degrades fibrin clots, which are responsible for most acute myocardial infarction cases (Castellino et al. 1976; Lee 1995). The plasmin can be obtained by the action of urokinase, tissue plasminogen activator, or by plasminogen-streptokinase complex (1:1). The complex is then transformed into streptokinase-plasmin-complex by fibrinolytic activity that converts the inactive proenzyme plasminogen to active plasmin (Hermel et al. 2010; Rashedi et al. 2024). The 230 amino acid residues at N-terminal of streptokinase share sequence similarity to trypsin-like serine proteases. However, streptokinase lacks a catalytic serine residue and does not display the characteristic properties of trypsin-like enzymes. Streptokinase lacks esterase or amidase activities against synthetic substrates and also incapable of cleaving peptide bonds in proteins. Streptokinase is also resistant to di-isopropyl fluorophosphate or pancreatic trypsin inhibitor (Reddy 1988).

Streptokinase is a plasminogen activator with extraordinary specificity by converting the enzymatically inactive plasminogen within the fibrinolytic system to the active plasmin with protease activity (Kunamneni et al. 2006). Due to this uniqueness, streptokinase differs from other thrombolytic agents, such as tissue plasminogen activator and urokinase. The active plasmin is capable of dissolving fibrin clots that is subsequently eliminated by phagocytes (Kunamneni et al. 2006) and hence thereby help in restoring the normal blood flow (Reddy 1988). For commercial application, streptokinase is formulated with the addition of human albumin as a stabilizing agent. The purifications of streptokinase have been achieved using ammonium sulfate fractionation, gel filtration, and ion-exchange chromatography (Reddy 1988). Experimental studies also revealed that the fibrin, fibrinogen, and fragment D can augment the activity of streptokinase. It is also reported that the presence of fibrinogen enhances the rate of active site formation within the streptokinase-plasminogen complex, without affecting its activator capabilities or the presence of fibrinogen is known to enhance the activator activity of the complex. Furthermore, fibrin exhibited higher stimulation compared to even fibrinogen in activating GLU-plasminogen when bound to streptokinase-plasminogen complex. These stimulatory factors also activated the amidolytic activity of the streptokinase-plasminogen complex for S-2251, synthetic substrate (Reddy 1988). The immobilization of streptokinase has been done using calcium alginate beads, agarose gel, and gelatin formaldehyde gel. It was observed that the maximum efficacy of $81.8 \pm 1.06\%$

was achieved using calcium alginate beads, followed by 71.0 ± 1.54% using cross-linked gelatin formaldehyde gel, and then 55.6 ± 2.17% when agarose gel was used as matrix (Vaishnavi and Subathra 2017).

Despite all these the significant advancements, the applications of streptokinase in the clinical applications is still full of challenges and limitations. The primary concern is its immunogenic nature which can lead to the production of antibodies that may diminish the efficacy of subsequent streptokinase therapy and increase the risk of adverse reactions, such as anaphylaxis.

4.2 Source, History, and Mechanisnistic Insight into Action of Streptokinase

Streptokinase, a non-enzymatic extracellular protein, is produced by various strains of beta-hemolytic streptococci. Several years of effort by William Smith Tillett and his student Sol Sherry led to the discovery of streptokinase as a thrombolytic agent for treating acute myocardial infarction. Streptokinase, obtained from the strains of hemolytic streptococci, is capable of transforming human plasminogen to plasmin (Tillett and Garner 1933). Streptokinase, a thrombolytic agent derived from the bacterium *Streptococcus*, has played a pivotal role in the treatment of various cardiovascular disorders. The action of streptokinase was initially believed to be a direct enzymatic action on fibrin. Therefore, initially, it was referred to as fibrinolysin, However, in 1941, it was observed that the effect of streptokinase in executed through the activation of a plasma protein. Later on, it was determined that it indirectly induces fibrinolysis by activating a plasma protein within the fibrinolytic system in humans. The term "streptokinase" was later coined by Christensen to describe the bacterial extract. The mechanism of streptokinase revealed its unique behavior as it lacks serine residue at its active site and does not exhibit the classical properties of trypsin-like enzymes. The enzyme binds to plasminogen and then converts into plasmin, which catalyze the dissolution of fibrin clots (Figs. 4.1 and 4.2) (Reddy 1988). Over the years, streptokinase has witnessed significant advancements, both in terms of its therapeutic applications and the challenges associated with its development and administration (Moussaddy et al. 2018; Che et al. 2020). Streptokinase consists 414 amino acids residues that form a single-chain polypeptide of 47 kDa in size with an isoelectric pH of 4.7. The protein is known to exhibit optimum activity between pH 7.3 and 7.6. Unlike urokinase and tissue-type plasminogen activator, streptokinase does not possess any inherent proteolytic activity, and instead activates plasminogen to plasmin indirectly by forming a high-affinity equimolar complex with PG (Kunamneni et al. 2006; Rafipour et al. 2019).

The three groups, that is, A, C, and G of beta-hemolytic streptococcal bacteria that belong to the Lancefield groups are the primary sources of streptokinase. Among these, the strains from group C are preferred for streptokinase production. The reason for preferring group C is their inability to produce erythrogenic toxin. A group C strain of *Streptococcus equisimilis* has been extensively utilized for streptokinase production (Christensen 1945). The reason for the selection of strain was

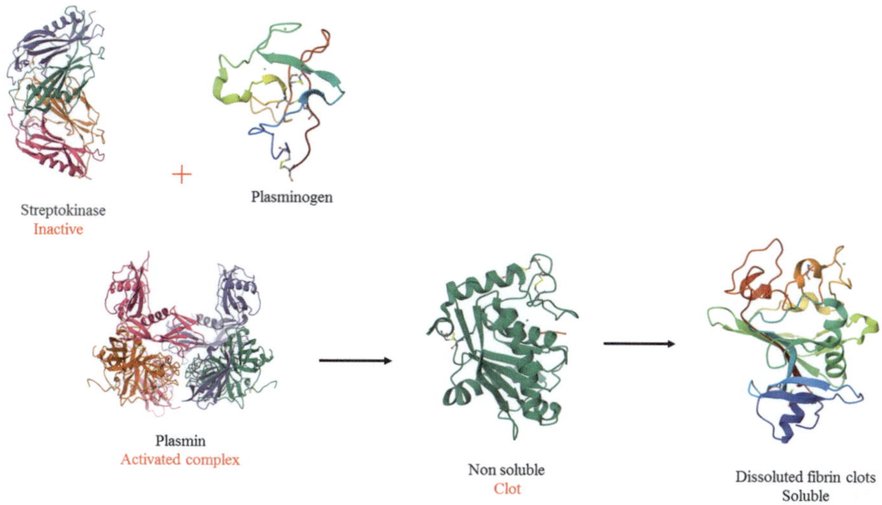

Fig. 4.1 Mechanism of Streptokinase in fibrinolysis

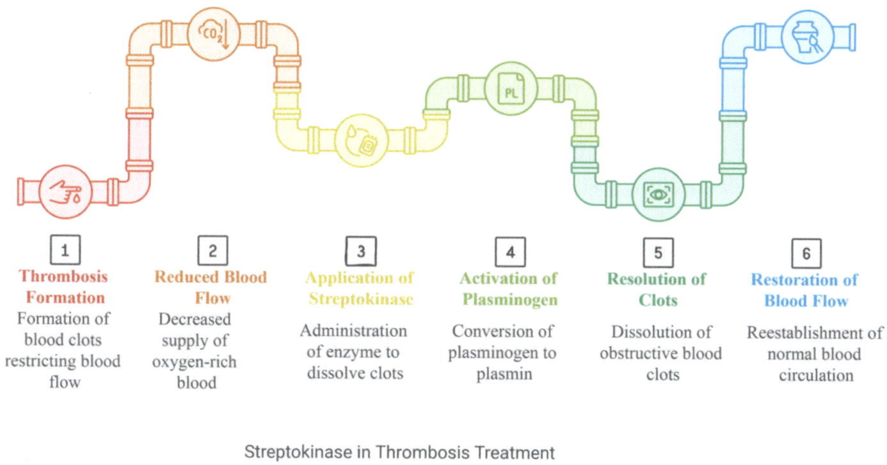

Fig. 4.2 Application of Streptokinase for thrombolysis

its inability to produce erythrogenic toxin and less fastidious growth requirements compared to the majority of group A strains that allowed the use of a semi-defined medium for the optimal growth and streptokinase secretion. Furthermore, this *Streptococcus equisimilis* strain has served as the principal source for the streptokinase gene, which has been employed to express the protein in *Escherichia coli*, *Streptococcus*, and yeast (Kunamneni et al. 2006).

The history of streptokinase in pharmaceutical applications dates back to the 1950s, when it was first discovered as a potent activator of the fibrinolytic system (Bergmann and Hammerschmidt 2007). The early study revealed that the application of a combination of streptokinase and plasminogen or plasma significantly enhanced the fibrinolysin-pathogenic staphylococcal strain for mice compared to streptokinase and plasminogen when administered alone (Ri and Young 1959). The clinical potential of streptokinase as a thrombolytic agent in fibrinolytic system was realized in the 1950s, when Tillett and Johnson demonstrated its effectiveness in the treatment of thrombosis in rabbits. The local applications of sodium morrhuate induced artificial intravascular clots within the ear vein of the rabbits. The intravenous infusion of streptokinase (at 40,000 units per kilo per hour) into the opposite ear liquefied and disappeared these clots in rabbits (Johnson and Tillett 1952). Since then, streptokinase has been extensively studied and utilized in the management of severe acute respiratory distress syndrome (ARDS) (Abdelaal Ahmed Mahmoud et al. 2020), percutaneous catheter drainage (PCD) (Bhargava et al. 2022), and others such as thrombotic conditions, including acute myocardial infarction, deep vein thrombosis, and pulmonary embolism (Knuttinen et al. 2010).

The application of streptokinase at 250,000 IU after 4 h significantly decreased plateau pressure and enhanced the compliance at day 8, the only treatment that decreased $PaCO_2$ and also reduced the ICU mortality. The streptokinase treatment enhanced the PaO_2/F_IO_2 ratio (arterial pO_2/fraction (percent) of inspired oxygen also known as P/F ratio) from >100 to >200 to >300 at day 7 one first, third, and fifth day (Abdelaal Ahmed Mahmoud et al. 2020). In PCDs patients, the necrosum can lead to blockage, and the application of streptokinase (at 50,000 IU) or H_2O_2 (at 3%) in saline irrigation decreased the requirement of necrosectomy and rate of mortality (Bhargava et al. 2022). The early trials of streptokinase for the treatment of ischemic stroke were discouraging, with negative results or increased risk of death. However, in 1995, a clinical study conducted by the National Institute of Neurological Disorders and Stroke demonstrated that patients treated with recombinant tissue plasminogen activator within 3 h of symptom onset achieved better neurological recovery and experienced less disability than those who received a placebo. The continuous development and the comparative studies on of recombinant streptokinase have further expanded its potential applications, offering a more cost-effective and standardized approach to thrombolytic therapy (Hernández et al. 1992; Diwedi et al. 2005; Hernández-Bernal et al. 2015; Leng and Xiong 2019).

The role of streptokinase includes treating invasive diseases such as necrotizing soft tissue infections. *Streptococcus dysgalactiae* subspecies *equisimilis* has emerged as a significant cause of invasive diseases, including necrotizing soft tissue infections. The diversity of microbial agents can cause these infections. The necrotizing soft tissue infections are predominantly attributed to *Streptococcus pyogenes*. Necrotizing soft tissue infections are uncommon yet severe clinical conditions that cause the necrosis of skin layer and other soft tissue structures. These infections frequently exhibit systemic toxicity, necessitating intensive care and extensive surgical interventions for the majority of patients. Even with optimal management, necrotizing soft tissue infections are associated with high rates of morbidity and mortality. The microscopic investigations revealed that the bacterial aggregations cause biofilm formation in SDSE-associated necrotizing soft tissue infections. Furthermore, genetic manipulation studies have demonstrated that the deletion of the streptokinase encoding gene in the SDSE S118 strain enhanced the capacity for biofilm production. Interestingly, the streptokinase-deficient mutant strain exhibited elevated metabolic activity, and subsequent profiling of the biofilm revealed a higher accumulation of a diverse array of metabolites, compared to the wild-type strain (Tölken et al. 2024).

4.3 Purification and Characterization of Streptokinase

The streptokinase has a long history for its use as a thrombolytic agent. Besides this, the key issues associated with its application include its antigenic nature. The other major concerns include its short half-life, which demands the administration of high dose that further enhances risk of adverse effects. To address these challenges, efforts have been made to improve the purification and formulation of streptokinase. So far strains of *Streptococcus* such H46A, that is, American Type Culture Collection No.12449 and Rebecca C. Lancefield strain H46A, 1956, have been extensively used for the isolation, characterization, and cloning of corresponding genes. An extracellular protein it has been isolated from hemolytic streptococcus group C (Avilán et al. 1997; Cherish Babu et al. 2008). The purification studies revealed that native streptokinase consists of single chain polypeptide of 414/415 amino acid residues with a molecular mass of 47 kDa in size and isoelectric pH of 4.7. Upon activation, the protein is known to exhibit optimum activity between pH 7.3 and 7.6 (Kunamneni et al. 2006; Rafipour et al. 2019). Streptokinase consists of 69 carboxyl side chains, 61 basic side chains, and 170 hydrophobic residues. The protein lacks cysteine or cystine amino acid residues and therefore lacks disulphide bond and glycosylation. The clinical applications of streptokinase include its application for the treatment of acute myocardial infarction, arterial thrombosis and embolism, deep vein thrombosis, and acute retinal vein occlusion.

The continuous research efforts led to the evolution of streptokinase for commercial applications. The distribution of streptokinase genetic diversity has been studied in 134 different strains containing 61 of M types, 6 of T types as well as in other streptococcal species using corresponding encoding gene of group C streptococcal strain H46A as a probe in colony and Southern hybridization. The presence of streptokinase gene was observed only in the pathogenic group belonging to A, C, and G streptococci, whereas it was absent in 12 belonging to Lancefield group strains. The casein-plasminogen overlay assay revealed that out of 143, only 83 exhibited streptokinase activity. The restriction digestion using *Hind* III and *Hae* III and Southern hybridization revealed that the bands highly conserved site for *Hind* III (Huang et al. 1989).

The purification of native streptokinase can be achieved through techniques such as ammonium sulfate fractionation, gel filtration, and ion-exchange chromatography. These purification methods have helped to reduce the presence of impurities and improve the overall quality of streptokinase formulations. The recombinant streptokinase has been obtained after expression in *E. coli* and *Pichia pastoris* as well as using cell free system. The recombinant expression of streptokinase for therapeutic purpose has been done using HeLa-based cell free protein system (CFPS system), 6× His, or GST. In general, the presence of tag such as 6× His or GST has been reported to facilitate the purification of proteins, but these tags have been reported for affecting the efficacy of streptokinase (86 percent compared to 52.5 and 43.1 for HIS and GST tagged, respectively) (Tran et al. 2018). In addition to this, the proteins such as human albumin has been added as a stabilizer in preparations of streptokinase formulation. The streptokinase-encoding gene has been successfully cloned from the *Streptococcus equisimilis* H46A strain and expressed in *E. coli* (Malke and Ferretti 1984). Additionally, research on streptokinase gene across diverse sources revealed polymorphism in its structure. Furthermore, the streptokinase alleles have been reported from 47 isolates of *S. pyogenes*. The streptokinase-encoding gene (*skc*) from the *S. equisimilis* H46A strain has been successfully expressed in various heterologous gram-positive and gram-negative bacteria. A high-level periplasmic and extracellular expression of streptokinase of 5000 IU/ml has been obtained using plasmid pSK100 (Ko et al. 1995). Additionally, the *skc* has been expressed in eukaryotic host such as yeast *Pichia pastoris* (Kunamneni et al. 2006).

The accumulation of streptokinase causes toxicity to the host cells and hence the optimization of conditions for large-scale production of streptokinase are challenging. Therefore, the large-scale production of streptokinase is often achieved as inclusion bodies. The inclusion bodies are then solubilized to using detergents of ionic, non-ionic nature, and salts and then refolded to get an active protein. Other challenges associated with streptokinase include tissue targeting and peripheral bleeding. Ongoing research aims to address these limitations through the development of novel delivery systems and formulation strategies.

4.4 Applications and Concerns for Streptokinase

Pharmacological interventions may include the administration of aspirin tablets and the injection of thrombolytic or clot-dissolving drugs, such as tissue plasminogen activator, streptokinase, or urokinase, within 3 h of the onset of a heart attack. Painkillers, such as morphine or meperidine, may also be used to alleviate pain (Lü et al. 2015). The history of streptokinase as a thrombolytic agent dates back to the 1950s (Che et al. 2020). At global level, thrombo-occlusion, that is, the formation of blood clot is the leading cause of morbidity and mortality. Therefore, therapeutic measures involving the revascularization and restoration of blood flow are critical to ensure the tissue survival and ultimately the health of patient. The clinical applications of streptokinase involve targeting the acute myocardial infarction, arterial thrombosis and embolism, deep vein thrombosis, and acute retinal vein occlusion (Cross and Derdeyn 2014). As described earlier, the activity of streptokinase is manifested through the conversion of human plasminogen to active plasmin. The application of streptokinase for the treatment of acute myocardial infarction, or heart attacks was a pioneer initiative. The researchers in late 1950s performed groundbreaking experiments related to the intravenous application of streptokinase against acute stroke treatment (Fletcher et al. 1959).

The potential of streptokinase in Indian perspective can be realized from the fact that the national intervention council data revealed approximately 21,000 primary percutaneous coronary interventions till 2011. Considering India's current population of over 1.5 billion and the estimated burden of myocardial infarction, the pharmaco-invasive therapy utilizing fibrin-specific thrombolytics represents a viable alternative. Therefore, the pharmaco-invasive therapy utilizing fibrin-specific thrombolytics represents a viable alternative. However, in numerous countries, it is not the fibrin-specific fibrinolytic that is widely accessible; rather, streptokinase, which is not only widely available but also cost-effective, continues to be the prominent option, as evidenced by its persistent presence in the literature from less industrialized nations (Mishra 2017).

The production of streptokinase remains challenging due to the complexity of the fermentation process. Despite this, streptokinase is a crucial thrombolytic agent in the management of various cardiovascular disorders. Its value as a therapeutic agent and its evolution using genetic engineering and production optimization have led to its future pharmaceutical developments in the field of thrombosis prevention and treatment. Moreover, the streptokinase-mediated thrombolysis involves a complex interplay between the streptokinase of bacterial origin and the host's fibrinolytic system. It is particularly effective in treating acute ischemic events, where rapid dissolution of the clot is crucial for minimizing tissue damage and improving patient outcomes. Efforts also have been made by the researchers to prepare liposomal formulations that can enhance the thrombolytic potency of streptokinase. These liposomal preparations have the potential to improve plasma stability, prolong drug retention in circulation, and facilitate selective delivery to the thrombus. Additionally, the use of liposomal carriers can mitigate the side effects, such as immunogenicity and hemorrhage, associated with conventional streptokinase formulations. The therapeutic efficacy of liposomal streptokinase and the potential of different liposomal carriers to improve the thrombolytic efficacy of streptokinase have been well demonstrated in animal studies (Kumar et al. 2017).

Streptokinase activates human plasminogen to plasmin by forming a high-affinity equimolar complex with plasminogen. Severe acute respiratory distress syndrome is characterized by severe hypoxemia and multiple organ failures due to the inability to recruit widespread collapsed alveoli. The role of nebulized streptokinase and nebulized unfractionated heparin (10,000 IU/4 hours) lies in their potential use as rescue therapies for reversing alveolar collapse nonresponsive to conventional management. Both are well absorbed through the pulmonary capillary membrane, owing to the lung's features such as a large absorption area, extensive vascularization of the respiratory mucosa, and low enzymatic activity. A study on 1796 individuals revealed that the streptokinase group had more ventilator-free days compared to the heparin and placebo. Streptokinase is slowly absorbed into the systemic circulation through the alveoli similar to other large protein molecules like α1-antitrypsin and albumin. The gradual absorption of streptokinase through the alveoli is likely responsible for its pronounced effect in severe ARDS and hypoxemia, potentially by dissolving both the intra-alveolar and alveolar-capillary clots, thereby reversing the obstruction of both alveolar ventilation and perfusion (Lee 1995; Mahmoud et al. 2019).

4.5 Side Effects of Streptokinase

The major concern with the application of streptokinase is the risk of its immunogenic nature. Therefore, the exposure to it causes an immune response in some cases. Despite all challenges, the effectiveness of streptokinase in dissolving blood clots with potential applications in restoring blood supply and preventing the tissue has led to its emergence as a standard therapy for curing cardiovascular emergencies. Enzyme is recommended for the treatment of acute ischemic conditions, that demands the fast reperfusion to improve the patient health. The results of the initial application of streptokinase were disappointing due to high hemorrhagic risk. The relative short shelf-life of streptokinase demands the high dose of enzymes through continuous infusion that often requires a close surveillance of patients under treatment (Che et al. 2020). However, the continuous developments in scanning and a better insight of thrombolysis helped in gaining significant improvements in the applications of streptokinase-based therapies for acute stroke management (Cross and Derdeyn 2014). The pharmaceutical potential of streptokinase has been improved using recombinant engineering approaches. The expression and production of streptokinase has been explored using codon optimization and other gene manipulation to enhance the yields of recombinant streptokinase and also reduced the antigenic behavior (Che et al. 2020). Despite the challenges, the role of streptokinase in the treatment of various thrombotic disorders remains significant. As the prevalence of cardiovascular diseases continues to rise globally, the demand for effective thrombolytic agents like streptokinase is also growing. Additionally, the ongoing research and development in the field of thrombolytic therapy, including the optimization of streptokinase production and the exploration of novel delivery methods, hold promise for further improving the safety, efficacy, and accessibility of this important pharmaceutical agent. These improvements have led to a more cost-effective and standardized approach to streptokinase-based thrombolytic therapy, addressing some of the previous limitations associated with its production and availability. For example, using a combination of codon optimization, gene dosage, and process optimization, the high level of recombinant streptokinase was achieved (Che et al. 2020).

Although the incidence of streptokinase-induced anaphylaxis is low, the reaction can be severe, especially in patients experiencing myocardial infarction, where further hemodynamic instability may lead to catastrophic outcomes (Tisdale et al. 1989). Studies suggested an approximately 5% incidence of allergic reactions in patients receiving their first dose of streptokinase. The reactions are typically minor such as rash, transient fever, or chills. The experimental study has revealed increased antibody levels against streptokinase in 9 out of 189 patients without prior exposure to streptokinase (Lee 1995). A case report describes a patient with acute myocardial infarction experienced an anaphylactic reaction shortly after the initiation of intravenous streptokinase therapy. The patient became profoundly hypotensive and developed a rapidly spreading erythematous rash that covered most of the body. Mechanical ventilation and epinephrine administration were required to stabilize the patient's blood pressure, as dopamine and norepinephrine were found ineffective. Streptokinase-induced anaphylaxis is known to be mediated by immunoglobulin E, and affected individuals have demonstrated elevated serum IgE concentrations to streptokinase compared to those who do not develop this adverse reaction. Epinephrine is the recommended treatment for managing hypotension associated with anaphylaxis. However, limited evidence exists to support the routine prophylactic use of corticosteroids and/or antihistamines in patients scheduled to receive streptokinase. Streptokinase skin testing may be a useful and reliable method for identifying individuals at risk of streptokinase-induced anaphylaxis (Tisdale et al. 1989; Levine et al. 1995; Bhasker 2022). Streptokinase has also been associated with various bleeding complications, including intracranial bleeding, which is the most dreaded complication.

4.6 Conclusion and Future Perspectives

Streptokinase is a widely available and cost-effective thrombolytic agent with a long history of clinical use. While it is not as potent as the newer fibrin-specific thrombolytics, it remains an important option, especially in resource-limited settings. Continued research into optimizing streptokinase therapy, including exploring novel delivery methods and combinations with other agents, will be crucial in expanding its clinical utility and improving patient outcomes. Recent advancements in drug delivery systems, such as liposomal formulations, have the potential to enhance the safety and efficacy of streptokinase by mitigating its side effects. In conclusion, the history and development of streptokinase in pharmaceutical applications have been marked by both successes and challenges. While the initial use of streptokinase was hampered by limitations such as non-selectivity and bleeding complications, the advancement of thrombolytic research has led to the optimization of streptokinase and the emergence of newer alternatives. Reflecting the ongoing research and development efforts, the pharmaceutical applications of streptokinase continue to evolve, with a focus on improving its safety, efficacy, and accessibility for the management of thrombotic disorders.

References

Abdelaal Ahmed Mahmoud A, Mahmoud HE, Mahran MA, Khaled M (2020) Streptokinase versus unfractionated heparin nebulization in patients with severe Acute Respiratory Distress Syndrome (ARDS): a randomized controlled trial with observational controls. J Cardiothorac Vasc Anesth 34(2):436–443. https://doi.org/10.1053/j.jvca.2019.05.035. Epub 2019 May 27

Avilán L, Yarzábal A, Jürgensen C, Bastidas M, Cruz J, Puig J (1997) Cloning, expression and purification of recombinant streptokinase: partial characterization of the protein expressed

in *Escherichia coli*. Braz J Med Biol Res 30(12):1427–1430. https://doi.org/10.1590/s0100-879x1997001200007
Bergmann S, Hammerschmidt S (2007) Fibrinolysis and host response in bacterial infections. Thromb Haemost 98(09):512–520. https://doi.org/10.1160/th07-02-0117
Bhargava MV, Rana SS, Gorsi U, Kang M, Gupta R (2022) Assessing the efficacy and outcomes following irrigation with streptokinase versus hydrogen peroxide in necrotizing pancreatitis: a randomized pilot study. Dig Dis Sci 67(8):4146–4153. https://doi.org/10.1007/s10620-021-07201-0. Epub 2021 Aug 17
Bhasker B (2022) P1621: anaphylactic shock to a hematological disorder patient during transfusion of a platelet apheresis unit stored in platelet additive solution: a rare case report. Hemasphere 6:1502–1503. https://doi.org/10.1097/01.hs9.0000849340.43505.66
Castellino FJ, Sodetz JM, Brockway WJ, Siefring GE Jr (1976) Streptokinase. Methods Enzymol 45:244–257. https://doi.org/10.1016/s0076-6879(76)45024-3
Che Z, Cao X, Chen G, Liang Z (2020) An effective combination of codon optimization, gene dosage, and process optimization for high-level production of fibrinolytic enzyme in *Komagataella phaffii* (*Pichia pastoris*). BMC Biotechnol 20(1). https://doi.org/10.1186/s12896-020-00654-7
Cherish Babu PV, Srinivas VK, Krishna Mohan V, Krishna E (2008) Renaturation, purification and characterization of streptokinase expressed as inclusion body in recombinant *E. coli*. J Chromatogr B Analyt Technol Biomed Life Sci 861(2):218–226. https://doi.org/10.1016/j.jchromb.2007.10.008. Epub 2007 Oct 13
Christensen LR (1945) Streptococcal fibrinolysis: a proteolytic reaction due to a serum enzyme activated by streptococcal fibrinolysin. J Gen Physiol 28:363–383
Cross, D T., & Derdeyn, C P. (2014). Therapeutic neuroradiology; intra-arterial thrombolysis. Elsevier BV, 433-436. doi:https://doi.org/10.1016/b978-0-12-385157-4.00206-2
Diwedi SK, Hiremath JS, Kerkar PG, Reddy KN, Manjunath CN, Ramesh SS, Prabhavati S, Dhobe M, Singh K, Bhusari P, Rao R (2005) Indigenous recombinant streptokinase vs natural streptokinase in acute myocardial infarction patients: phase III multicentric randomized double blind trial. Indian J Med Sci 59(5):200–207
Fletcher A, Sherry S, Alkjaersig N, Smyrniotis F, Jick S (1959) The maintenance of a sustained thrombolytic state in man. II. Clinical observations on patients with myocardial infarction and other thromboembolic disorders. J Clin Invest. 38(7):1111–1119. https://doi.org/10.1172/JCI103887. PMID: 13664786; PMCID: PMC293258
Gupta PJ (2007) Suppositories in anal disorders: a review. Eur Rev Med Pharmacol Sci 11:165–170
Hermel M, Dailey W, Hartzer MK (2010) Efficacy of plasmin, microplasmin, and streptokinase-plasmin complex for the in vitro degradation of fibronectin and laminin- implications for vitreoretinal surgery. Curr Eye Res 35(5):419–424. https://doi.org/10.3109/02713680903572517
Hernández L, Rodriguez P, Serrano R, Muñoz E, Estrada MP, Dela Fuente J, Herrera L (1992) Recombinant streptokinase for the treatment of thrombotic disorders. Ann N Y Acad Sci 667:424–427. https://doi.org/10.1111/j.1749-6632.1992.tb51643.x
Hernández-Bernal F, Castellanos-Sierra G, Valenzuela-Silva CM, Catasús-Álvarez KM, Martínez-Serrano O, Lazo-Diago OC, Bermúdez-Badell CH, Causa-García JR, Domínguez-Suárez JE (2015) THERESA-4 (Treatment of HEmorrhoids with recombinant streptokinase application) Group of Investigators. Recombinant streptokinase vs hydrocortisone suppositories in acute hemorrhoids: a randomized controlled trial. World J Gastroenterol 21(23):7305–7312. https://doi.org/10.3748/wjg.v21.i23.7305. PMID: 26109819; PMCID: PMC4476894
Hirsh J (1970) Dosage regimens for streptokinase treatment: evaluation of a standard dosage schedule. Australas Ann Med 19(s1):12–16. https://doi.org/10.1111/imj.1970.19.s1.12
Holt B, Gupta AS.2012). Streptokinase loading in liposomes for vascular targeted nanomedicine applications: encapsulation efficiency and effects of processing. J Biomater Appl 26(5):509–527. doi:https://doi.org/10.1177/0885328210374778. Epub 2010 Jul 21
Huang TT, Malke H, Ferretti JJ (1989) Heterogeneity of the streptokinase gene in group a streptococci. Infect Immun 57(2):502–506. https://doi.org/10.1128/iai.57.2.502-506.1989. PMID: 2643572; PMCID: PMC313124

Johnson AJ, Tillett WS (1952) The lysis in rabbits of intravascular blood clots by the streptococcal fibrinolytic system (streptokinase). J Exp Med 95(5):449–464. https://doi.org/10.1084/jem.95.5.449. PMID: 14927798; PMCID: PMC2212084

Khan F, Tritschler T, Kahn SR, Rodger MA (2021) Venous thromboembolism. Lancet 398(10294):64–77. https://doi.org/10.1016/S0140-6736(20)32658-1. Epub 2021 May 10

Knuttinen MG, Emmanuel N, Isa F, Rogers A, Gaba RC, Bui JT, Owens C (2010) Review of pharmacology and physiology in thrombolysis interventions. Semin Intervent Radiol 27(04):374–383. https://doi.org/10.1055/s-0030-1267850

Ko JH, Park DK, Kim IC et al (1995) High-level expression and secretion of streptokinase in *Escherichia coli*. Biotechnol Lett 17:1019–1024. https://doi.org/10.1007/BF00143093

Kumar L, Verma S, Vaidya B (2017) Liposomes for the delivery of streptokinase. Ther Deliv 8:855

Kunamneni A, Abdelghani TTA, Ellaiah P (2006) Streptokinase—the drug of choice for thrombolytic therapy. J Thromb Thrombolysis 23(1):9–23. https://doi.org/10.1007/s11239-006-9011-x

Lee HS (1995) How safe is the Readministration of streptokinase? Drug Saf 13(2):76–80. https://doi.org/10.2165/00002018-199513020-00002

Leng T, Xiong Z (2019) Treatment for ischemic stroke: from thrombolysis to thrombectomy and remaining challenges. Medknow 5(1):8–8. https://doi.org/10.4103/bc.bc_36_18

Levine M, Goldhaber SZ, Gore JM, Hirsh J, Califf RM (1995) Hemorrhagic complications of thrombolytic therapy in the treatment of myocardial infarction and venous thromboembolism. Chest 108(4):291S–301S. https://doi.org/10.1378/chest.108.4_supplement.291s

Lü L, Liu M, Sun R, Zheng Y, Zhang P (2015) Myocardial infarction: symptoms and treatments. Cell Biochem Biophys 72(3):865–867. https://doi.org/10.1007/s12013-015-0553-4

Mahmoud AAA, Mahmoud HE, Mahran M, Khaled MH (2019) Streptokinase versus unfractionated heparin nebulization in patients with severe acute respiratory distress syndrome (ARDS): a randomized controlled trial with observational controls. J Cardiothorac Vasc Anesth 34(2):436–443. https://doi.org/10.1053/j.jvca.2019.05.035

Malke H, Ferretti JJ (1984) Streptokinase: cloning, expression, and excretion by Escherichia coli. Proc Natl Acad Sci USA 81(11):3557–3561. https://doi.org/10.1073/pnas.81.11.3557. PMID: 6374659; PMCID: PMC345548

Mishra S (2017) What is the ideal fibrinolysis to PCI time: pharmaco-invasive strategy with streptokinase? Indian Heart J 69(5):569–570. https://doi.org/10.1016/j.ihj.2017.07.022

Moussaddy A, Demchuk AM, Hill MD (2018) Thrombolytic therapies for ischemic stroke: triumphs and future challenges. Neuropharmacology 134:272–279. https://doi.org/10.1016/j.neuropharm.2017.11.010

Ojha N, Dhamoon AS (2023) Myocardial infarction. In: StatPearls [internet]. StatPearls Publishing, Treasure Island

Rafipour M, Keramati M, Aslani MM, Arashkia A, Roohvand F (2019) The β-domain of streptokinase affects several functionalities, including specific/proteolytic activity kinetics. FEBS Open Bio 9(7):1259–1269. https://doi.org/10.1002/2211-5463.12657

Rashedi S, Greason CM, Sadeghipour P, Talasaz AH, O'Donoghue ML, Jimenez D, Monreal M, Anderson CD, Elkind MSV, Kreuziger LMB, Lang IM, Goldhaber SZ, Konstantinides SV, Piazza G, Krumholz HM, Braunwald E, Bikdeli B (2024) Fibrinolytic agents in thromboembolic diseases: historical perspectives and approved indications. Semin Thromb Hemost. 50(5):773–789. https://doi.org/10.1055/s-0044-1781451.Epub

Reddy K (1988) Streptokinase – biochemistry and clinical application. Enzyme 40(2–3):79–89. https://doi.org/10.1159/000469149

Ri K, Young G (1959) The streptokinase-plasminogen system. I. Its effect on the pathogenicity of streptococci and other organisms for mice. J Exp Med 110(2):245–258. https://doi.org/10.1084/jem.110.2.245. PMID: 13673137; PMCID: PMC2136989

Shen M, Wang Y, Hu F, Lv L, Chen K, Xing G (2021) Thrombolytic agents: nanocarriers in targeted release. Molecules 26(22):6776. https://doi.org/10.3390/molecules26226776. PMID: 34833868; PMCID: PMC8619279

Sherry (1969) The New England Journal of medicine. World Scientific:265–273. https://doi.org/10.1142/9789814273596_0013

Singh R, Gautam P, Sharma C, Osmolovskiy A (2023) Fibrin and fibrinolytic enzyme Cascade in thrombosis: unravelling the role. Life (Basel) 13(11):2196. https://doi.org/10.3390/life13112196. PMID: 38004336; PMCID: PMC10672518

Tillett WS, Garner RL (1933) The fibrinolytic activity of hemolytic streptococci. J Exp Med 58(4):485–502. https://doi.org/10.1084/jem.58.4.485. PMID: 19870210; PMCID: PMC2132310

Tisdale JE, Stringer KA, Antalek M, Matthews G (1989) Streptokinase-Induced Anaphylaxis. DICP 23(12):984–987. https://doi.org/10.1177/106002808902301206

Tölken LA, Neufend JV, Oppegaard O, Methling K, Moll K, Redanz S, Katsburg M, Ali MQ, Shumba P, Kreikemeyer B, Skrede S, Fulde M, Norrby-Teglund A, Lalk M, Kittang BR, Siemens N (2024) Streptokinase reduces *streptococcus dysgalactiae* subsp. *equisimilis* biofilm formation. BMC Microbiol 24(1):378. https://doi.org/10.1186/s12866-024-03540-w

Tran K, Gurramkonda C, Cooper MA, Pilli M, Taris JE, Selock N, Han TC, Tolosa M, Zuber A, Peñalber-Johnstone C, Dinkins C, Pezeshk N, Kostov Y, Frey DD, Tolosa L, Wood DW, Rao G (2018) Cell-free production of a therapeutic protein: expression, purification, and characterization of recombinant streptokinase using a CHO lysate. Biotechnol Bioeng 115(1):92–102. https://doi.org/10.1002/bit.26439. Epub 2017 Oct 6

Vaishnavi B, Subathra DC (2017) Potential application of immobilized streptokinase extracted from Streptococcus equinus VIT_VB2. Prep Biochem Biotechnol 47(10):1032–1036. https://doi.org/10.1080/10826068.2017.1373291. Epub 2017 Nov 20

Wang S, Reed GL, Hedstrom L (1999) Deletion of Ile1 changes the mechanism of streptokinase: evidence for the molecular sexuality hypothesis. Biochemistry 38(16):5232–5240. https://doi.org/10.1021/bi981915h

Therapeutics Applications of Nattokinase and Collagenases

5

Vivek Sharma and Sneh Sharma

Abstract

Collagenases and nattokinase are two distinct enzymes with significant therapeutic applications. Collagenases, a family of enzymes that break down collagen, are crucial for normal development, wound healing, and tissue regeneration. They are a broad family of proteins that make up approximately 30% of total protein composition and are a major protein in the extracellular matrix. Collagenases are used to treat conditions such as Peyronie's disease, wounds, burns, and glaucoma, as they can break down excess or abnormal collagen. Plant-derived collagenases are less utilized compared to microbial ones due to the complexity of plant systems. Nattokinase is a serine protease from fermented soybeans that exhibits powerful fibrinolytic (clot-dissolving) properties. It directly hydrolyzes fibrin and enhances the body's natural fibrinolytic activity by converting prourokinase to urokinase and elevating tissue plasminogen activator (t-PA) levels. Unlike conventional agents, nattokinase has minimal side effects and can be absorbed orally. Its therapeutic uses include reducing nasal polyps in chronic rhinosinusitis, potentially degrading the SARS-CoV-2 spike protein, and demonstrating antitumor effects by modulating the tumor microenvironment. It also has protective effects against retinopathy and exhibits antiplatelet and antithrombotic properties.

Keywords

Collagenase · Metalloproteases · Fibrinolytic enzyme · Glaucoma · Chronic rhinosinusitis · Antitumor · Anticoagulant

V. Sharma (✉) · S. Sharma
Department of Biotechnology, College of Horticulture and Forestry, Dr. Y.S. Parmar University of Horticulture and Forestry Solan, Hamirpur, Himachal Pradesh, India

© The Author(s), under exclusive license to Springer Nature Singapore Pte Ltd. 2026

V. Sharma, R. Salwan (eds.), *Pharmaceutically Relevant Microbial Enzymes*,
https://doi.org/10.1007/978-981-95-4941-2_5

5.1 Introduction

Collagen, a broad family of proteins, accounts for approximately 30% of total protein composition and acts as the major protein in the extracellular matrix (ECM) (Birk and Bruckner 2005). Its primary structure is made up of Gly, Ala, Pro and Hyp, whereas Cys, Met, Tyr, and His are low (Bae et al. 2008). Three polypeptide chains make up the common domain of all collagen molecules, which fold into a distinctive triple-helical shape (Boudko and Bächinger 2016). Only in extremely basic, highly acidic, and high-temperature environments can collagen be hydrolyzed, potentially destroying amino acids. Nonetheless, collagenase has a high selectivity and no detrimental impact on amino acids when hydrolyzing collagen under mild circumstances. As a result, it finds extensive usage in tissue engineering, medicinal therapy, and food processing (Wang et al. 2022). The enzyme collagenase is well known for its capacity to break down collagen, a protein that is widely present in both microbes and animals. Under typical physiological circumstances, it can break down native fibrillar collagen (Wang et al. 2022). Ines Mandl of Columbia University highlighted the significance of collagen as a key component of the skin, tendons, cartilage, teeth, and bones in an augration presentation at the first Collagenase Interdisciplinary Symposium. Normal development, wound healing, and tissue regeneration all depend on the breakdown and resynthesis of collagen fibers, which is facilitated by collagenase. Collagenase also has the ability to separate various cell types from dense connective tissue, including bones, tendons, and skin. The integrity of the cells and molecules is guaranteed by this enzymatic hydrolysis method, which is also more accurate and delicate than mechanical approaches. It is also essential for further research (Mandl 1970). The most widely studied collagenases are matrix metalloproteinases (MMPs) from animals and bacterial collagenases from micro-organisms. Zinc ions are present in the multi-domain proteolytic enzyme family known as MMPs (Wu et al. 2023; Sekhon 2010). Collagenases are indispensable in research because they allow scientists to explore the dynamics of collagen degradation and extracellular matrix remodeling, which are central to many biological processes. They are routinely employed for providing reliable material for cellular, molecular, and biochemical investigations. In biomedical research, collagenases serve as key tools for understanding pathological conditions such as fibrosis, arthritis, cancer metastasis, and impaired wound healing. Their controlled use also aids in developing experimental models for tissue engineering, regenerative medicine, and drug testing (Alipour et al. 2016). Thus, collagenases play a critical role in advancing both fundamental knowledge and translational research.

5.2 Sources of Collagenase

5.2.1 Plant Collagenases

A 30 kDa enzyme exhibiting protease and collagenase activity was extracted from crude ginger root powder. GP2, a ginger protease, can break down natural type I collagen at the same sites on all three chains, unlike traditional collagenases that focus on particular sequences (Kim et al. 2007). Proteases obtained from fig latex demonstrate significant specificity for gelatin and collagen, along with remarkable stability across different temperature and pH conditions, thereby possessing potential for various applications (Table 5.1). The use of collagenases derived from plants

Table 5.1 Classification of collagenases

Classification	Type of collagenase	Source	Applications
Microbial collagenases	Bacillus collagenases	*Bacillus subtilis* and related species	Moderate collagen-degrading ability; used in industry and biotechnology
	Clostridial collagenases (ColG, ColH, ColA)	*Clostridium histolyticum*	Highly active against native fibrillar collagen; applied in wound healing, tissue dissociation, and medical research.
	Vibrio collagenases	Marine bacteria	Aid in pathogenicity and environmental nutrient cycling
Plant-based proteases with collagen-degrading activity	Papain, bromelain	Papaya (*Carica papaya*), Pineapple (*Ananas comosus*)	Although not true collagenases, these enzymes can degrade collagen and are extensively applied in pharmaceuticals, food processing, and wound debridement.
Collagen-degrading enzymes in vertebrates (MMPs)	MMP-1 (Collagenase-1)	Mammals (fibroblasts, keratinocytes)	Capable of hydrolyzing collagens of type I, II, and III within connective tissues.
	MMP-8 (Collagenase-2)	Human neutrophils	An enzyme specifically involved in inflammation and immune regulation.
	MMP-13 (Collagenase-3)	Chondrocytes, osteoblasts	Demonstrates potent degradation of type II collagen, with key involvement in bone and cartilage remodeling and arthritis.
	MMP-18 (Collagenase-4	Amphibians (*Xenopus laevis*)	A developmental enzyme that contributes to tissue remodeling during growth.
Collagen-degrading enzymes in invertebrates	Collagenases from jellyfish, sea anemones	Marine invertebrates	Aid in predation and the digestion of prey tissues rich in collagen.

is presently restricted in comparison to that of its microbial counterparts due to the complex characteristics of plant systems (Raskovic et al. 2014). The utilization of plant-derived collagenases is currently limited compared with that of its microbial counterparts because of the intricate nature of plant systems (Wang et al. 2024).

5.2.2 Animal Collagenases

Widely found in vertebrates, animal collagenases show a clear substrate preference by cleaving the collagen triple helix at around three-quarters of its molecular length from the X-Gly bond on the α chain. Collagenolytic activity is shared by animal collagenases and other members of the zinc metallopeptidase family, including MMP-1 (interstitial collagenase), MMP-8 (neutrophil collagenase), MMP-13 (collagenase-3), and MMP-18 (Xenopus collagenase-4). In order to stabilize its tertiary structure, Zn 2+ and Ca 2+ ions are present (Lovejoy et al. 1994). In the peptidase database, animal collagenases, which belong to the broader family of MMPs, are categorized under the M10A subfamily. These zinc-dependent proteases, which are frequently called "matrixins," are essential for the turnover and remodeling of the extracellular matrix. In order to become a fully functional enzyme, collagenase must first be activated through particular proteolytic cleavage events. The "cysteine switch" is the process by which stromal proenzymes of the M10A subfamily are rendered inactive (Van Wart and Birkedal-Hansen 1990).

5.2.3 Microbial Collagenase

Nowadays, *C. histolyticum* fermentation is the main method used to create commercial collagenase. Based on differences in amino acid sequences and catalytic activities, microbial collagenases, which are metallopeptidases and members of the peptidase family M9, can be further divided into the M9A and M9B subfamilies Compared to plant and animal collagenases, collagenases produced from *Clostridium* and *Vibrio* have been studied in greater detail.

Collagenases, one of the most effective enzymes currently known for breaking down native collagen, are used by Clostridium to aid in host colonization and nutrition intake. The metalloproteinase family includes the huge multidomain proteins known as Clostridium collagenases (Eckhard et al. 2013).

5.3 Therapeutic Application of Collagenase

5.3.1 Peyronie's Disease

Five percent of males worldwide suffer from Peyronie's disease, sometimes referred to as induration penis plastica, a connective tissue ailment that causes fibrous plaques to form in the soft tissue of the penis (Pescatori 2015; Kadioglu et al. 2015).

The condition is caused by the deposition of too much collagen, which causes the penis to tilt to one side, get injured, and cease to function correctly. However, there are some treatments for this disease. Collagenase *C. histolyticum*, often known as Xiaflex, is an FDA-approved therapy for the treatment of Dupuytren's contracture and as an injectable medicine for Peyronie's disease. The medicine is thought to operate by breaking down the excess collagen in the penis that causes Peyronie's disease (Clark and Peter 2013).

5.3.2 Wound Healing

Experimental studies have demonstrated that collagenase enzymes in the wound healing process can promote the proliferation, angiogenesis, and migration of dermal cells (Eldor et al. 2004). Collagenase enzymes kill the life process of necrotic tissues in the wound area while causing minimal damage to healthy tissue. However, the development of granulation and epithelial cells is critical. Sometimes the healing process for bedsores and colloidal scars is not completed adequately, and the wound site becomes an ideal environment for the growth of infectious germs. In these cases, collagen levels must be decreased and regulated. However, this is difficult to achieve since collagen is resistant to proteases. In 2004, both the USA FDA and the UK Prescription Pricing Authority recognized the value of *Lucilia sericata* larvae in the treatment of wounds, and sterile maggots can now be formally prescribed (Gur et al. 2011).

5.3.3 Burns

The standard treatment of burned tissues by collagenase has already been approved by the FDA (McCallon et al. 2015). According to a study, collagenase ointment promotes faster wound healing than conventional burn care. In one study, 41 patients whose tissues surrounding burn wounds were surgically removed were compared to 78 burn patients treated with collagenase *C. histolyticum* ointment (Rashaan et al. 2014). In partial-layer burns, collagenase reduces the total requirement for surgery and blood transfusions while also allowing for a brief hospital stay. Therefore, in children with a partial-thickness burn wound that is not infected, collagenase may be the first therapy option to erase scarring (Benz et al. 2010).

5.3.4 Glaucoma

Glaucoma disease is the leading cause of permanent blindness. This condition affected around 60.5 million individuals in 2010, with an anticipated increase to 79.6 million by 2020. Therapies that lower intraocular pressure are commonly used to treat glaucoma, regardless of the type. Collagenase injection is a way of treating glaucoma (Alipour et al. 2016). The use of collagenase in drug manufacture aids in the treatment of intraocular scarring and fibrosis that occurs during glaucoma filtration surgery and the implantation of set-on implants. Glaucoma occurs when the eye's fluid drainage canals constrict due to aberrant collagen formation. This condition accounts for over 40% of all cases of blindness. Using collagenase to treat glaucoma was patented in 2013 (Alipour et al. 2016).

5.3.5 Vitrectomy

Collagenase is also used in certain vitrectomy situations to eliminate fibroproliferative tissue. It has been found that after 10 min, 15 min, and 30 min of incubation, collagenase that has been adequately purified can cause substantial degradation of scar tissue. The retina's internal limiting membrane and scar cellular sections have not shown any signs of phenotypic damage.

5.3.6 Nattokinase

Natto, a staple in the Japanese diet for more than 2000 years, is believed to play a significant role in the country's reputation for longevity and good health (Arafa et al. 2022; Nagata et al. 2016). In 1980, Japanese researcher Hiroyuki Sumi working at the Chicago University Medical School discovered that natto possess the ability to break down artificial fibrin, a key component in blood clots (Sumi et al. 1987). Sumi and his colleagues successfully isolated an enzyme from natto that demonstrated the ability to degrade both fibrin and a plasmin- specific substrate. This newly identified fibrinolytic enzyme was subsequently named "nattokinase" (Sumi et al. 1987). As the key thrombolytic ingredient of natto, the serine protease nattolkinase (NK) is proposed to be responsible for many of the benefits of natto on cardiovascular system, including thrombosis, hypertension, atherosclerosis, acute ischemic stroke, and hyperlipidemia (Wu et al. 2023a). Additionally, nattokinase (NK) has exhibited beneficial effects in alleviating Alzheimer's disease, various other disorders and certain forms of cancer. Growing evidence suggest that nattokinase offers protection against these age-related conditions, primarily due to its proteolytic activity (Wu et al. 2023a). This fibrinolytic enzyme with significant thrombolytic potential is synthesized by *Bacillus subtilis* during the fermentation of soybeans in the traditional Japanese food, Natto. Extensive research on Nattokinase has been conducted in East Asian countries, particularly Japan, Korea, and China. In recent years, its clot-dissolving efficacy has attracted growing attention and recognition within western medicine (Weng et al. 2017). Research focusing on standardized production methods, pharmacokinetics, clinical efficacy, safety evaluation, and targeted delivery systems will provide a strong foundation for integrating nattokinase into mainstream healthcare.

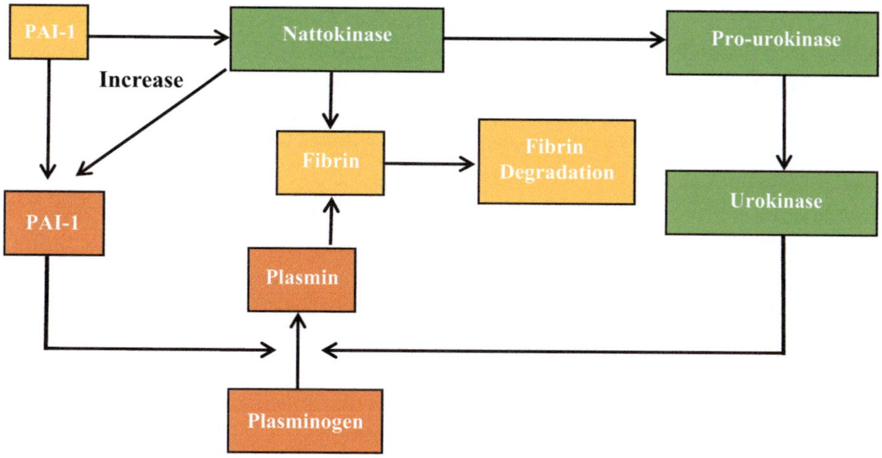

Fig. 5.1 Thrombolytic action of nattokinase

5.3.7 Mechanism of Thrombolytic Action of Nattokinase

Nattokinase facilitates the breakdown of blood clots through multiple mechanisms: it directly hydrolyzes fibrin and plasmon-specific substrates, promotes the conversion of endogenous prourokinase to its active form (uPA-1), degrades plasminogen activator inhibitor-1 (PAI-1), and enhances the levels of tissue plasminogen activator (t-PA), thereby supporting and amplifying fibronolytic activity (Fig. 5.1). Mechanistic insight of nattokinase) (Yatagai et al. 2008). In contrast to conventional fibrinolytic agents like tissues plasminogen activator (t-PA) and urokinase (upA), which are often associated with adverse effects such as bleeding complications, nattokinase has been reported to exhibit minimal to no side effects. Furthermore, research suggests that nattokinase is capable of being absorbed through the intestinal tract following oral administration, supporting its potential as a safe and non-invasive therapeutic option for thrombolytic therapy (Fujita et al. 1995, 2011). Nattokinase promotes thrombolysis through multiple mechanisms; it directly hydrolyzes fibrin and plasmin substrates, facilitates the conversion if endogenous prourokinase into its active form, urokinase (uPA), degrades plasminogen activator inhibitor (PAI-1), and elevates the concentration of tissue plasminogen activator (t-PA). Collectively these actions enhance the body's natural fibrolytic activity (Weng et al. 2017).

5.3.8 Therapeutic Applications of Nattokinase

Chronic Rhinosinusitis with Nasal Polyps (CRSwNP)

It is a frequently encountered condition, with a notably higher incidence in individuals aged over 65 years (Merrill and Kanaan 2022). Recent investigations have demonstrated that nattokinase (NK) contributes to a deduction in the size of nasal polyps in CRSwNP patients, primarily through its fibrinolytic activity. Moreover, NK has been shown to decrease the viscosity of nasal secretions in CRSwNP and sputum in individuals with asthma (Takabayashi et al. 2017).

SARS-Cov-2

A recent study reported that nattokinase (NK) is capable of degrading the spike (S) protein, a critical component required for viral entry into host cells. This finding indicates that NK may possess inhibitory potential against SARS-CoV-2 through the targeted degradation of the S protein (Tanikawa et al. 2022).

Antitumor Effect

Tumor development is driven by complex molecular mechanisms, including oncogenes, genetic mutations, and environmental influence. The tumor physical micro environment (TPME) plays a crucial role in regulating tumor cell proliferation, metabolism, metastasis, and treatment response. The intratumoral administration of nattokinase in solid tumors has been demonstrated to positively modulate TPME, thereby enhancing oxygen within tumor and drug accumulation (Zhang et al. 2023). Furthermore, Zhang et al. (2023) demonstrated that pretreatment with nattokinase prior to chimeric antigen receptor (CAR) T cell therapy significantly suppressed tumor progression. Similarly, Yan et al. (2019) reported that treatment with crude nattokinase extract in a heptocellular carcinoma (HCC) mouse model led to reduced expression of tumor-associated markers and prolonged survival.

Antiretinopathy Effect

Retinopathy can develop from several causes, including diabetic retinopathy (DR) driven by chronic hyperglycemia, renal vascular diseases associated with hypertension or arteriosclerosis, and retinal vein occlusion (Al-Shabrawey et al. 2013; Huang et al. 2021). Maintaining the structural and functional integrity of the blood retinal barrier (BRB) is critical in managing diabetic retinopathy. Huang et al. (2023) demonstrated that nattokinase can enhance BRB integrity by suppressing the expression of high-mobility group box 1 protein, thereby interrupting detrimental downstream signaling pathways.

Anticoagulant Effect

A comparative study evaluating the antiplatelet properties of nattokinase and aspirin revealed that nattokinase exhibited strong antiplatelet aggregation and antithrombotic effects both in vitro and in vivo. Specially< NK was found to suppress the formation of thromboxane B2 in collagen- stimulated platelets (Jang et al. 2013). Additionally, nattokinase demonstrated beneficial hemorheological effects in vitro by reducing red blood cell aggregation and lowering blood viscosity under low-shear conditions (Weng et al. 2017).

5.4 Conclusion

Both collagenases and nattokinase are powerful enzymes with distinct yet significant therapeutic applications. Collagenases, particularly MMPs, are vital for processes like wound healing and tissue regeneration due to their ability to break down collagen, a key component of the extracellular matrix. Their use has been approved for treating conditions such as Peyronie's disease, burns, chronic wounds, and glaucoma. While animal and microbial sources of collagenase are widely used, the application of plant-derived collagenases is currently limited. Nattokinase, a serine protease derived from the fermented soybean food, Natto, is a potent fibrinolytic agent. It works by directly hydrolyzing fibrin and enhancing the body's natural clot-dissolving mechanisms. Unlike traditional thrombolytic drugs, nattokinase is associated with minimal side effects and can be absorbed orally, making it a promising non-invasive therapy. Research has highlighted its efficacy in treating chronic rhinosinusitis with nasal polyps, and it has shown potential in combating SARS-CoV-2 by degrading the spike protein. Furthermore, its ability to modulate the tumor microenvironment suggests a role in cancer therapy, and it has demonstrated protective effects against retinopathy and potent antiplatelet properties. In essence, the therapeutic potential of both collagenases and nattokinase underscores the growing importance of enzyme-based therapies in modern medicine.

Conflict of Interest The authors declare that there is no conflict of interest and the images in this chapter are self-drawn.

References

Alipour H, Raz A, Zakeri S, Dinparast DN (2016) Therapeutic applications of collagenase (metalloproteases): a review. Asian Pac J Trop Biomed 6(11):975–981

Al-Shabrawey M, Elsherbiny M, Nussbaum J, Othman A, Megyerdi S, Tawfik A (2013) Targeting neovascularization in ischemic retinopathy: recent advances. Expert Rev Ophthalmol 8(3):267–286. https://doi.org/10.1586/eop.13.17

Arafa A, Kokubo Y, Kashima R, Teramoto M, Sakai Y, Nosaka S, Nakao YM, Watanabe E (2022) The lifelong health support 10: a Japanese prescription for a long and healthy life. Environ Health Prev Med 27(0):23–23. https://doi.org/10.1265/ehpm.22-00085

Bae I, Osatomi K, Yoshida A, Osako K, Yamaguchi A, Hara K (2008) Biochemical properties of acid-soluble collagens extracted from the skins of underutilised fishes. Food Chem 108(1):49–54

Benz MS, Packo KH, Gonzalez V, Pakola S, Bezner D, Haller JA et al (2010) A placebo-controlled trial of microplasmin intravitreous injection to facilitate posterior vitreous detachment before vitrectomy. Ophthalmology 117(4):791–797

Birk DE, Bruckner P. Collagen suprastructures. In: Brinckmann J, Notbohm H and Muller PK. 2005. Collagen: primer in structure, processing and assembly. Berlin\Heidelberg: Springer. 185–205.

Boudko SP, Bächinger HP (2016) Structural insight for chain selection and stagger control in collagen. Sci Rep 6(1):37831

Clark R, Peter H (2013) The molecular and cellular biology of wound repair. Springer, Gewerbestrasse

Eckhard U, Schönauer E, Brandstetter H (2013) Structural basis for activity regulation and substrate preference of Clostridial collagenases G, H, and T. J Biol Chem 288(28):20184–20194
Eldor R, Raz I, Ben Yehuda A, Boulton A (2004) New and experimental approaches to treatment of diabetic foot ulcers: a comprehensive review of emerging treatment strategies. Diabet Med 21(11):1161–1173
Fujita M, Hong K, Ito Y, Misawa S, Takeuchi N, Kariya K, Nishimuro S (1995) Transport of nattokinase across the rat intestinal tract. Biol Pharm Bull 18(9):1194–1196. https://doi.org/10.1248/bpb.18.1194
Fujita M, Ohnishi K, Takaoka S, Ogasawara K, Fukuyama R, Nakamuta H (2011) Antihypertensive effects of continuous oral administration of nattokinase and its fragments in spontaneously hypertensive rats. Biol Pharm Bull 34(11):1696–1701. https://doi.org/10.1248/bpb.34.1696
Gur S, Limin M, Hellstrom WJ (2011) Current status and new developments in Peyronie's disease: medical, minimally invasive and surgical treatment options. Expert Opin Pharmacother 12(6):931–944
Huang Z, Ng TK, Chen W, Sun X, Huang D, Zheng D, Yi J, Xu Y, Xi Z, Chen S-L (2021) Nattokinase attenuates retinal neovascularization via modulation of Nrf2/HO-1 and glial activation. Invest Ophthalmol Vis Sci 62(6):25. https://doi.org/10.1167/iovs.62.6.25
Huang Z, Chu WK, Ng TK, Chen S, Liang J, Chen C, Xu Y, Xie B, Ke S, Liu Q, Chen W, Huang D (2023) Protective effects of nattokinase against microvasculopathy and neuroinflammation in diabetic retinopathy. J Diabetes 15(10):866–880. https://doi.org/10.1111/1753-0407.13439
Jang JY, Kim TS, Cai J, Kim J, Kim Y, Shin K, Kim KS, Park SK, Lee SP, Choi E-K, Rhee MH, Kim YB (2013) Nattokinase improves blood flow by inhibiting platelet aggregation and thrombus formation. Lab Anim Res 29(4):221–221. https://doi.org/10.5625/lar.2013.29.4.221
Kadioglu A, Boyuk A, Salabas E (2015) Re: clinical efficacy of collagenase *Clostridium histolyticum* in the treatment of Peyronie's disease by subgroups: results from two large, double-blind, randomized, placebo-controlled, phase III studies. Eur Urol 68(5):908–909
Kim M, Hamilton SE, Guddat LW, Overall CM (2007) Plant collagenase unique collaenolytic activity of cysteine proteases from ginger. Biochim Biophys Acta 1770(12):1627–1635
Lovejoy B, Cleasby A, Hassell AM, Longley K, Luther MA, Weigl D, McGeehan G, McElroy AB, Drewry D, Lambert MH, Jordan SR (1994) Structure of the catalytic domain of fibroblast collagenase complexed with an inhibitor. Science 263(5145):375–377
Mandl I (1970) Collagenase. Science 169(3951):1234–1238
McCallon SK, Weir D, Lantis JC 2nd. (2015) Optimizing wound bed preparation with collagenase enzymatic debridement. J Am Coll Clin Wound Spec 6(1–2):14–23
Merrill T, Kanaan A (2022) Managing chronic rhinosinusitis with nasal polyps in the elderly: challenges and solutions. Clin Interv Aging 17:685–698. https://doi.org/10.2147/cia.s279765
Nagata C, Wada K, Tamura T, Konishi K, Goto Y, Koda S, Kawachi T, Tsuji M, Nakamura K (2016) Dietary soy and natto intake and cardiovascular disease mortality in Japanese adults: the Takayama study. Am J Clin Nutr 105(2):426–431. https://doi.org/10.3945/ajcn.116.137281
Rashaan ZM, Krijnen P, Klamer RR, Schipper IB, Dekkers OM, Breederveld RS (2014) Nonsilver treatment *vs.* silver sulfadiazine in treatment of partial-thickness burn wounds in children: a systematic review and meta-analysis. Wound Repair Regen 22(4):473–482
Raskovic B, Bozovic O, Prodanovic R, Niketic V, Polovic N (2014) Identification, purification and characterization of a novel collagenolytic serine protease from fig (*Ficus carica* var.Brown Turkey) latex. J Biosci Bioeng 118(6):622–627
Pescatori ES (2015) Surgical treatment of Peyronie's disease by inflatable penile prosthesis. In: Cavallini G, Paulis G (eds) Peyronie's disease. Springer, Gewerbestrasse, pp 141–147
Sekhon BS (2010) Matrix metalloproteinases an overview. Res Rep Biol:1–20
Sumi H, Hamada H, Tsushima H, Mihara H, Muraki H (1987) A novel fibrinolytic enzyme (nattokinase) in the vegetable cheese Natto; a typical and popular soybean food in the Japanese diet. Experientia 43(10):1110–1111. https://doi.org/10.1007/bf01956052
Takabayashi T, Imoto Y, Sakashita M, Kato Y, Tokunaga T, Yoshida K, Narita N, Ishizuka T, Fujieda S (2017) Nattokinase, profibrinolytic enzyme, effectively shrinks the nasal polyp tis-

sue and decreases viscosity of mucus. Allergol Int 66(4):594–602. https://doi.org/10.1016/j.alit.2017.03.007

Tanikawa T, Kiba Y, Yu J, Hsu K, Chen S, Ishii A, Yokogawa T, Suzuki R, Inoue Y, Kitamura M (2022) Degradative effect of Nattokinase on spike protein of SARS-CoV-2. Molecules 27(17):5405. https://doi.org/10.3390/molecules27175405

Van Wart HE, Birkedal-Hansen H (1990) The cysteine switch a principle of regulation of metalloproteinase activity with potential applicability to the entire matrix metalloproteinase gene family. Proc Natl Acad Sci 87(14):5578–5582

Wang Y, Wang P, Cao HY, Ding HT, Su HN, Liu SC, Liu G, Zhang X, Li CY, Peng M (2022) Structure of Vibrio collagenase VhaC provides insight into the mechanism of bacterial collagenolysis. Nat Commun 13(1):566

Wang ZZ, Wang K, Xu LF, Su C, Gong JS, Shi JS, Ma XD, Xie N, Qian JY (2024) Unlocking the potential of collagenase: structure, function, and emerging therapeutic horizons. BioDesign Res 6:0050. https://doi.org/10.34133/bdr.0050

Weng Y, Yao J, Sparks S, Wang KY (2017) Nattokinase: an oral antithrombotic agent for the prevention of cardiovascular disease. Int J Mol Sci 18(3):523. https://doi.org/10.3390/ijms18030523

Wu S, Zhou X, Jin Z, Cheng H (2023) Collagenases and their inhibitors a review. Collagen Leather 5(1):19

Wu H, Zhang Q, Suo H, Xu F, Huang W, Wang DO (2023a) Nattokinase as a functional food ingredient: therapeutic applications and mechanisms in age-related diseases. Food Sci Human Wellness 13(5):2401–2409. https://doi.org/10.26599/fshw.2022.9250198

Yan Y, Wang Y, Qian J, Wu S, Ji Y, Liu Y, Zeng J, Gong A (2019) Nattokinase crude extract inhibits hepatocellular carcinoma growth in mice. J Microbiol Biotechnol 29(8):1281–1287. https://doi.org/10.4014/jmb.1812.12058

Yatagai C, Maruyama M, Kawahara T, Sumi H (2008) Nattokinase-promoted tissue plasminogen activator release from human cells. Pathophysiol Haemost Thromb 36(5):227–232. https://doi.org/10.1159/000252817

Zhang Y, Pei P, Zhou H, Xie Y, Yang S, Shen W, Hu L, Zhang Y, Liu T, Yang K (2023) Nattokinase-mediated regulation of tumor physical microenvironment to enhance chemotherapy, radiotherapy, and CAR-T therapy of solid tumor. ACS Nano 17(8):7475–7486. https://doi.org/10.1021/acsnano.2c12463

Microbial Sacrosidase: Mechanism of Action and Its Therapeutic Applications

6

Tuyelee Das and Priyanka Sood

Abstract

Sacrosidase (invertase) is a glycoside hydrolase that catalyzes the hydrolysis of sucrose into glucose and fructose. Sacrosidase is naturally produced in the human small intestine. In contrast, microbial sacrosidase primarily from yeasts (e.g., *Saccharomyces cerevisiae*), fungi (e.g., *Aspergillus niger*), and certain bacteria offers scalable production through fermentation for both industrial and therapeutic applications. Initially valued in the food and beverage industry for invert sugar production, sacrosidase has evolved into a clinically relevant enzyme for treating congenital sucrase-isomaltase deficiency (CSID). Production strategies include submerged fermentation (SmF) and solid-state fermentation (SSF), with SSF often yielding higher enzyme titers for specific strains. Advances in statistical optimization methods, recombinant DNA technology, and genetic engineering have significantly enhanced enzyme yield, purity, and stability. Downstream processes such as membrane filtration and immobilization further improve industrial feasibility by increasing operational stability and reusability. Invertase also exhibits several health benefits, with emerging applications in functional foods, nutraceuticals, and biopharmaceuticals. This chapter integrates biochemical and applied perspectives, highlighting the transformation of microbial sacrosidase from a historical biochemical curiosity into a versatile, high-value biocatalyst with enduring significance in both health and industry.

T. Das · P. Sood (✉)
University Centre for Research Development (UCRD), Chandigarh University, Mohali, Punjab, India

© The Author(s), under exclusive license to Springer Nature Singapore Pte Ltd. 2026

V. Sharma, R. Salwan (eds.), *Pharmaceutically Relevant Microbial Enzymes*,
https://doi.org/10.1007/978-981-95-4941-2_6

Keywords

Microbial sacrosidase · Invertase · β-fructofuranosidases

6.1 Introduction

Sacrosidase (trade name sucraid) is a medicine that replaces sucrase in those who lack this enzyme. It is accessible as an oral solution (Treem et al. 1999). In humans, sacrosidase is naturally produced by the brush-border membrane of the small intestine and plays a crucial role in carbohydrate digestion. A deficiency in this enzyme results in congenital sucrase-isomaltase deficiency (CSID), a rare genetic disorder characterized by malabsorption, osmotic diarrhea, and related gastrointestinal symptoms (Treem 2012). Sacrosidase assists in the breakdown of sugar/sucrose into simpler forms and is useful for the relief of gastrointestinal symptoms that are associated with CSID. Microbial sacrosidase refers to the enzyme obtained from microbial sources, particularly yeasts (e.g., *Saccharomyces cerevisiae*), fungi (e.g., *Aspergillus niger*), and certain bacteria, which can be produced at scale using fermentation technologies (Nadeem et al. 2015). The first microbial sacrosidase was identified in yeast extracts during early studies on sugar fermentation in the nineteenth century, leading to the broader understanding of enzymatic catalysis by Eduard Buchner in 1897 (Buchner 1897).

Subsequent advancements in microbiology and industrial biotechnology enabled large-scale production of invertase from *S. cerevisiae*, initially for confectionery and beverage industries. Later, medical research identified its therapeutic potential, particularly for treating sucrase deficiencies. This marked a substantial shift in sacrosidase's perception as both an industrial catalyst and a clinically useful therapeutic enzyme. Invertase or β-fructofuranosidases (β-D-fructofuranoside fructohydrolase EC 3.2.1.26) is the common word for sacrosidase used in applications within food and industry (Veana et al. 2018). Additionally, Sucraid® refers explicitly to highly purified sacrosidase also from baker's yeast, made according to pharmaceutical standards for the treatment of CSID. Invertases are glycoside hydrolases which initiate the catalysis of the cleavage of sucrose or sugar or α-D-glucopyranosyl-β-D-fructofuranoside. These further divided into two monosaccharides, i.e., glucose and fructose (Henrissat and Bairoch 1996; Naumoff 2001; Sturm and Tang 1999).

Recombinant DNA technology further enabled the expression of sacrosidase in engineered microbial systems presenting improved purity, yield, and stability. The supplementation of sacrosidase has been proven therapy for supplementation of sucrase deficiency disease enabling the restoration of normal carbohydrate digestion and resultant improvement in quality of life. Microbial sacrosidase finds applications in the food and beverage industry for producing invert sugar, in bioethanol production, and as a processing aid in pharmaceuticals. Recent research has also explored its role in functional foods, prebiotic formulations, and targeted metabolic therapies. This chapter focuses on microbial sacrosidase, emphasizing its enzymatic mechanism, structural biology, and kinetic properties, along with its therapeutic applications in gastrointestinal disorders and metabolic regulation. Additionally, it will explore industrial-scale production strategies, recombinant approaches, and potential emerging uses in nutraceutical and biopharmaceutical sectors. The discussion aims to integrate fundamental biochemical understanding with applied perspectives to highlight how microbial sacrosidase bridges human health and industrial biotechnology.

6.2 Enzyme Classification

Invertases are included in glycoside hydrolase (GH) family 32 based on their amino acid sequences homology (Lombard et al. 2013). Invertases are known by several names. It is also referred to as β-fructofuranosidase (EC 3.2.1.26), β-D-fructofuranoside fructohydrolase, sucrose glycosidase, fructosidase, sucrase, invertin, maxinvert, β-fructosylinvertase, glucosucrase, saccharase, and sucrase (Manoochehri et al. 2020; Timerman 2012). Invertases are broadly categorized based on their optimal pH levels and their location within the cell. Based on their optimal pH, invertases are divided into two main groups. Acid invertases function best in acidic environments, with an optimal pH ranging from 4.5 to 5.5 (Barbosa et al. 2018; Liu et al. 2025; Osiebe et al. 2023). Neutral/alkaline invertases are most active in neutral to alkaline conditions, with an optimal pH between 7.0 and 7.8 (Du et al. 2010; Nadeem et al. 2015; Sturm et al. 1999; Zhou et al. 2016). Microbial invertases are majorly classified as acid invertases, while alkaline invertases form a distinct class with physicochemical properties that differ from acid invertases (Zhou et al. 2016; Manoochehri et al. 2020). Bacterial invertases can be active over a wide pH range, and their optimal pH levels can differ (Du et al. 2013; Yoon et al. 2007). They also differ in the presence or absence of carbohydrate groups in their protein structure. Extracellular invertases usually lack these groups, whereas intracellular invertases have them (Lincoln and More 2017).

Invertases also exist as various isoforms, which are distinguished by their specific locations within the cell. These isoforms can be either intracellular (within the cell) or extracellular (outside the cell) (Barbosa et al. 2018; Manoochehri et al. 2020). Invertases from *Aspergillus niger* (Rubio and Maldonado 1995) and *A. ochraceus* (Guimaraes et al. 2007) are homodimeric proteins with molecular masses of 95 kDa and 135 kDa, respectively. In *A. nidulans*, both high- and low-molecular-weight isoforms (S- and F-forms) that differ in dimerization and glycosylation were reported (Chen et al. 1996). Another study by Wolska-Mitaszko et al. (2007) reported the presence of two secreted isoforms with molecular weights of 60 and 120 kDa in *Fusarium oxysporum*. Among yeast strains, *S. cerevisiae* produces more levels of two forms of invertase, periplasmic glycosylated form and cytosolic non-glycosylated form (Belcarz et al. 2002; Lincoln and More 2017). While the study by Andjelković et al. (2010) reported four isoforms of extracellular invertase from *S. cerevisiae* with no difference in the structure of the protein and have similar catalytic properties. In another study, the presence of native invertase as either a dimeric enzyme (205 kDa) or a multimeric complex (1070–1200 kDa) was reported in

Schizosaccharomyces pombe (Moreno et al. 1990). Moreover, invertases from *Arxula adeninivorans* (Boer et al. 2004) and *Rhodotorula glutinis* (Rubio et al. 2002) exist as hexameric (600 kDa) and dimeric (100 kDa) enzymes, respectively.

6.3 Mechanism of Action of Sacrosidase

β-fructofuranosidases or invertases (β-D-fructofuranoside fructohydrolase EC 3.2.1.26), hydrolyze β-fructosyl bonds located on the non-reducing ends of various β-D-fructofuranoside substrates (sucrose) to release glucose and fructose (Nascimento et al. 2016; Linde et al. 2009). Under optimal conditions, these types of enzymes can also show transfructosylation activity (Vo et al. 2021) to synthesize short-chain fructo-oligosaccharides (FOS) of various lengths. The FOS can consist of 1–3 fructosyl units that are joined as an attachment to a sugar backbone, and the type of glycosidic linkage that is formed will vary depending on the source of the enzyme (Antosova and Polakovic 2001; Sangeetha et al. 2005). The common FOS are 1-kestose (β-D-fructofuranosyl-(2 → 1)-β-D-fructofuranosyl-(2 → 1)-α-D-glucopyranoside), nystose (β-D-fructofuranosyl-(2 → 1)-β-D-fructofuranosyl-(2 → 1)-β-D-fructofuranosyl-(2 → 1)-α-D-glucopyranoside), and fructosyl-nystose (Fernandes 2025). Furthermore, transfructosylation may also form other products such as levan and inverted syrups (Ozimek et al. 2006). This enzyme exhibits both hydrolytic and transfructosylating activities as well as broad substrate specificity. It catalyzes the cleavage of the β-(1 → 2) glycosidic bond of a broad range of sugars (i.e., sucrose, kestose, raffinose, nystose, leucrose, and inulin) and transfers the fructosyl unit to another sugar molecule (Álvaro-Benito et al. 2007; Gutiérrez-Alonso et al. 2009).

The mechanism of the production of FOS from sugars, such as sucrose, by β-fructofuranosidases proceeds via either reverse hydrolysis or transfructosylation (Antosova and Polakovic 2001). Reverse hydrolysis is an equilibrium process where low water activity drives the reaction toward synthesis of FOS rather than the hydrolytic breakdown of sucrose.

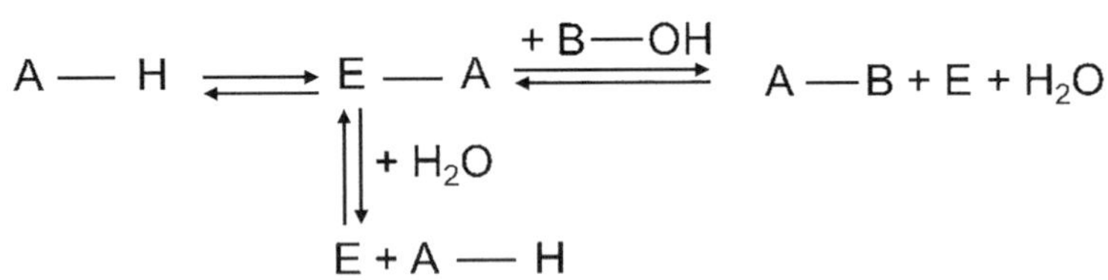

Where, A represents the fructosyl group of the donor molecule, B is the glycosyl group of the acceptor (such as sucrose or an existing FOS), and E denotes the enzyme β-fructofuranosidase. The enzyme first forms an intermediate complex (E–A), which then transfers the fructosyl group to the acceptor (B), resulting in the formation of a product A–B (FOS) that is one fructosyl unit longer than the original acceptor molecule.

In the initial step of the reaction, an activated enzyme–donor complex (E–A) is formed. This complex subsequently reacts with a hydroxyl group of an acceptor molecule. The acceptor can be either water, leading to hydrolysis or sucrose, resulting in reverse hydrolysis. The yield of FOS produced depends on the equilibrium constants governing the transformation of the E–A complex into either free enzyme and fructose, or kestose. Increasing the substrate concentration, and thereby reducing water activity, shifts the reaction equilibrium toward synthesis, ultimately enhancing the yield of FOS.

In case of transfructosylation, a fructosyl group is transferred from an activated donor, typically sucrose, to a suitable acceptor molecule. This reaction occurs in two steps, beginning with the formation of an activated donor–enzyme complex.

$$\begin{array}{ccc} A\text{—}B + E & \rightleftharpoons & B\text{—}OH + E\text{—}A & \xrightleftharpoons{+\,C\text{—}OH} & A\text{—}C + E + H_2O \\ & & \Big\updownarrow{+\,H_2O} & & \\ & & E + A\text{—}H & & \end{array}$$

Where, A–B represents the fructosyl donor (such as sucrose), B–OH is glucose, and C–OH is the acceptor molecule (sucrose or a FOS). E denotes the enzyme β-fructofuranosidase. The reaction results in the formation of A–C, a product that is a FOS extended by one fructosyl unit relative to the original acceptor.

Transfructosylation is a kinetically controlled process as the product (A–C) can also serve as a substrate for the enzyme β-fructofuranosidase. As a result, the concentration of FOS depends on the relative rates of synthesis and hydrolysis. Hirayama et al. (1989) reported that FOS levels initially increased, reached a plateau, and then declined—a trend particularly evident in dilute solutions and at high sucrose conversion. Therefore, high FOS yields are generally obtained by conducting reactions at elevated sucrose concentrations.

6.4 Microbial Sources of Sacrosidase

β-fructofuranosidases are found in a wide variety of organisms, including higher plants, animals, and microorganisms (Aleksanyan and Markosyan 1986; Nadeem et al. 2015; Shiomi 1992; Van den Ende and Van Laere 1996). For industrial-scale production microorganisms like filamentous fungi (*Aspergillus, Penicillium, Fusarium, Gloesporium*), yeasts (*Saccharomyces, Rhodotorulla, Pichia, Hansenula, Candida, and Aureobasidium*), and some bacteria are the viable and preferred sources (Álvaro-Benito et al. 2007; Lincoln and More 2017; Lorentz et al. 1995; Nascimento et al. 2016; Nguyen et al. 2005; Win et al. 2004). This preference is due to their high enzyme yields, ease of cultivation, amenability to genetic modification, and suitability and cost-effectiveness for large-scale processing.

These versatile enzymes are not uniformly distributed within the cell. They can be found in different cellular compartments (Barbosa et al. 2018; Lincoln and More 2018; Manoochehri et al. 2020). For example, in maize (*Zea mays*), the enzyme is located in the cell wall (Niu et al. 2020; https://www.uniprot.org/uniprotkb/P49174/entry). Furthermore, β-fructofuranosidases exist in various isoforms, which can be present as monomers or as more complex polymers (Ramírez-Escudero et al. 2016). For example, β-fructofuranosidase from *Xanthophyllomyces dendrorhous* exists as a homodimeric protein (Linde et al. 2009). Conversely, most biochemically characterized β-fructofuranosidases from *bifidobacteria* have been reported as monomeric (Jedrzejczak-Krzepkowska et al. 2011).

6.4.1 Filamentous Fungi

Historically and currently, filamentous fungi are the dominant source for commercially produced β-fructofuranosidases. Major fungal genera known to produce β-fructofuranosidases are *Aureobasidium* (Yoshikawa et al. 2006), *Penicillium* (Matei et al. 2017; Nascimento et al. 2016; Xu et al. 2015), *Aspergillus* (Batista et al. 2021; Han et al. 2025; Kurakake et al. 2008; Lorenzoni et al. 2014; Moreira et al. 2024; Mussatto et al. 2009; Punt et al. 2002), *Fusarium* (Gonçalves et al. 2013; Onitilo et al. 2022), *Rhodotorula* (Gutiérrez-Alonso et al. 2009), *Rhizopus* (Matei et al. 2017), etc.

Their widespread use is mainly due to their ability to secrete the enzyme extracellularly, which simplifies downstream purification. Additionally, fungal sources also possess high capacity for enzyme production in submerged fermentation systems (Manoochehri et al. 2020). Among fungi, species belonging to the genus *Aspergillus* hold significant importance. *Aspergillus niger* and *Aspergillus japonicus* have a long history of use and are recognized for their high productivity (Trollope et al. 2015; U.S. FDA 2018). Enzymes derived from these sources hold "Generally Recognized as Safe" (GRAS) status by regulatory bodies like the US FDA (Vo et al. 2021; Zorn et al. 2024). They are considered acceptable in food applications in many countries around the world, including Australia, New Zealand, Canada, and China (FSANZ 2021; Health Canada 2021; Legifrance 2006; MFDS 2019). Such approval from many regulatory authorities underlines their potential commercial significance. Other species such as *Aspergillus nidulans, A. oryzae, A. foetidus, A. thermomutatus, A. awamori*, *A. sydowi*, and *A. labruscus* have also been identified as potent producers (Chen et al. 2023; Maresma et al. 2010; Moreira et al. 2024; Ottoni et al. 2012; Smaali et al. 2012; Tódero et al. 2019).

β-Fructofuranosidases from various fungi exhibit dual functionality, performing both hydrolytic and transfructosylating activities (Fernandez et al. 2004). Under appropriate reaction conditions such as pH, temperature, and substrate concentration, the transfructosylating activity may be increased (Fernandez et al. 2004). Fructosyltransferases are well characterized and produce a variety of FOS that have prebiotic characteristics associated with health benefits (Belmonte-Izquierdo et al. 2023; Cuervo-Fernandez et al. 2007; Gutiérrez-Alonso et al. 2009). The immobilization of β-fructofuranosidases has shown to increase thermal stability and reuse of the enzymes, which can enhance the industrial potential (Lorenzoni et al. 2014). Fructosyltransferases from *Aspergillus* species are widely used in industry to produce FOS, which are a polymer of β-(2 → 1)-linked fructose oligomers. Along with *Aspergillus* as a source, *Aureobasidium pullulans* is an industrial source of enzymes specifically for the synthesis of FOS. The search for novel fungal strains with fructosyltransferase activity remains an ongoing research focus. Efforts are focused on identifying new isolates that possess high stability and strong catalytic activity. Other known genera that produce fructosyltransferases include *Ceratocystis, Penicillium, Fusarium, Cladosporium, Trichoderma,* and *Verticillium* (Araújo et al. 2023; Bhatti et al. 2006; Gonçalves et al. 2013; Matei et al. 2017; Nascimento et al. 2016).

Additionally, other important factors that influence enzyme yield and activity are the fermentation technology, solid-state (SSF) or submerged (SmF) fermentation, specified culture conditions, and choice of the microbial strains (Batista et al. 2021; Fernandes et al. 2018; Oliveira et al. 2020). *Aspergillus niger* has shown promising invertase production through SSF. Additionally, *A. niger* GH1, isolated from the Mexican semi-desert, was able to produce high levels of invertase yield (5231 U/L) using hydrolyzed sugarcane bagasse and hydrolyzed molasses as the substrates (Veana et al. 2014). In another study, optimization of culture conditions for *A. niger* using response surface methodology (RSM) resulted in maximum invertase production (8.2 U/ml) at pH 5, 35 °C and 72 h of incubation (Salim et al. 2020). *Saccharomyces cerevisiae* MK also demonstrated optimal invertase production at specific conditions of pH 6 and 30 °C (Shankar et al. 2013). Yet in another study, Osiebe et al. in (2023) evaluated intracellular invertase production by *S. cerevisiae* OL629078.1 while using cassava-soybean medium and yeast sucrose broth. The comparative results revealed 4.1-fold increase in enzyme activity on using cassava-soybean medium (318.605 U/mg) compared to the yeast sucrose broth (77.6 U/mg) (Osiebe et al. 2023). This highlights the potential of alternative, cost-effective substrates for enhancing enzyme production. Another factor affecting enzyme yield is genetic improvement through parasexual recombination as studied in *A. niger* strains. This approach significantly enhanced invertase productivity with diploid strains showing 5–18 times higher levels than haploid strains in SSF (Montiel-González et al. 2002).

Together these studies highlight the potential of optimized fungal strains and fermentation and culture conditions for industrial-scale invertase production. However, the production of biotechnologically important enzymes remains largely restricted to a limited number of fungal species. Exploring the enzymatic potential of other fungal strains is essential for identifying novel and valuable enzyme sources.

6.4.2 Yeasts: The Classic and Versatile Source

The use of yeast as a source of β-fructofuranosidases has progressed significantly over time. These advancements have expanded both the diversity and potential applications of the enzyme. Yeast β-fructofuranosidases have been widely studied, including *S. cerevisiae* (Farine et al. 2001; Goud et al. 2013), *Schizosaccharomyces pombe* (Tanaka et al. 2013), *Pichia anomala* (Pérez et al. 1996; Rodriguez et al. 1995), *Candida utilis* (Plascencia-Espinosa et al. 2014; Belcarz et al. 2002), or *Schwanniomyces occidentalis* (Rodrigo-Frutos et al. 2021). Yeast β-fructofuranosidases are also distinguished based on their ability to produce prebiotic FOS. One enzyme that stands out is from *Schwanniomyces occidentalis*, in that it produced high yields of the prebiotic 6-kestose (Álvaro-Benito et al. 2007). Early studies focused on *Kluyveromyces fragilis*, which produced inulase. The inulase produced is beneficial for commercial/industrial purposes compared to *S. cerevisiae* invertase (GrootWassink and Fleming 1980). *S. cerevisiae* produces both extra- and intracellular β-fructofuranosidases varying in molecular weights under optimized conditions (Rashad et al. 2006). Later, *Xanthophyllomyces dendrorhous* was identified as a producer of a novel β-fructofuranosidase (Xd-INV). This enzyme has unique properties, like high thermostability and extensive glycosylation with ability to synthesize neo-FOS including neokestose and neonystose (Linde et al. 2009). Gimeno-Pérez et al. (2015) focused on heterologous expression to improve enzyme production in a previous study, and they managed to successfully express the enzyme in *Pichia pastoris* and reported an almost 1500-fold increase in the levels of enzyme compared to any previous heterologous system. The enzyme activity was also nearly double compared to the natural producer while keeping most of its biochemical properties and ability to produce neo-FOS intact (Gimeno-Pérez et al. 2015). This heterologously expressed enzyme is the first reported yeast enzyme to produce blastose. These types of studies display various properties and potential uses of yeast β-fructofuranosidases in producing prebiotic sugars.

In another study, yeast β-fructofuranosidase from *Rhodotorula dairenensis* (RdINV) was characterized and heterologously expressed in *P. pastoris*. RdINV produces a mixture of levan-, inulin-, and neo-FOS, which is further characterized by having 6-kestose as the major product (Gimeno-Pérez et al. 2021). RdINV, like Xd-INV, is categorized as glycosyl hydrolase (GH) family 32. Combining the fructosyl transfer to produce a variety of products makes RdINV and Xd-INV interesting targets for industrial applications in the manufacturing of prebiotic sugars (Gimeno-Pérez et al. 2021; Linde et al. 2009). While *Saccharomyces* species remains as the primary industrial fermentative organism, research to examine the use of non-conventional yeasts is growing. These yeast species will be investigated for their ability to enhance product complexity and develop new enzymatic properties (Steensels and Verstrepen 2014). These advancements have expanded the range of yeast species used for β-fructofuranosidase production and improved enzyme yields for industrial applications.

6.4.3 Bacteria: An Emerging and Diverse Frontier

Although fungi and yeasts have traditionally dominated the field, bacteria represent a vast and largely unexplored source of β-fructofuranosidases with distinctive characteristics. Bacterial invertases exhibit diversity in glycosylation, polymerization, and phosphorylation and are produced by both gram-positive and gram-negative bacteria (Antosova and Polakovic 2001; Siddiqa and Amin 2022). Bacteria can produce intracellular and extracellular invertases. *Bacillus cereus* TA-11, an alkalophilic and thermophilic bacterium, produces an intracellular invertase with optimal activity at pH 7.0 and 60 °C (Yoon et al. 2007). *Zymomonas mobilis* also produces an intracellular invertase, whose gene has been cloned and sequenced, revealing similarities with other β-D-fructofuranosidases (Yanase et al. 1991). *Streptomyces* sp. ALKC8, an actinomycete, produces extracellular invertase with optimal activity at 60 °C and pH 5.0 (Kaur and Sharma 2005). Invertase production can be optimized through various factors, including medium composition and incubation conditions (Kaur and Sharma 2005). These diverse bacterial invertases have potential applications in pharmaceuticals, analytical probes, and food industries (Siddiqa and Amin 2022).

However, studies on bacterial β-fructofuranosidases are very limited; thus, identifying novel strains of bacteria with the potential to produce β-fructofuranosidase is required. Extensive studies have shown that *Bifidobacterium* spp., categorized as GRAS (generally regarded as safe), have the potential to produce β-fructofuranosidase (Chen et al. 2023; Muramatsu et al. 1992). These include *B. longum* (Ávila-Fernández et al. 2016), *B. adolescentis* (Muramatsu et al. 1992), *B. breve* (Ryan et al. 2005), *B. lactis* (Janer et al. 2004), and *B. infantis* (Warchol et al. 2002). Other aspects of these β-fructofuranosidases have also been studied, including enzyme production (Lincoln and More 2017; Schallmey et al. 2004), enzyme purification (O'Mullan et al. 1992), biochemical characterization (Kaur and Sharma 2005; Lincoln and More 2017; Yoon et al. 2007), and the cloning and expression of β-fructofuranosidase genes (Chen et al. 2014; Kurimoto et al. 1999). These preliminary exploratory studies have shown that this type of β-fructofuranosidase has good prospects for use in various industries, including the production of invert sugar. However, these studies have all focused on *Bifidobacterium* species and their genes for producing β-fructofuranosidase. The hydrolyzation of raffinose and stachyose using these characterized β-fructofuranosidases from *Bifidobacterium* spp. has rarely been useful because of their limited resources.

6.5 Production Methods

Several key elements must be addressed to achieve larger enzyme titers and increase the synthesis of numerous primary and secondary metabolites, including the selection of an appropriate microbe and accurate control of process parameters. The integration of industrial by-products into fermentation media provides a low-cost alternative carbon source while also addressing waste management issues. Further modification of culture parameters can considerably improve fermentation efficiency.

Enzyme production is generally carried out using solid-state fermentation (SSF), semi-solid fermentation, or submerged fermentation (SmF). In recent years, SmF has gained considerable global adoption for invertase production due to its high productivity per reactor volume, cost-effectiveness, and suitability for precise culture control and downstream processing. SmF is particularly advantageous in processes where crude fermented products can be used directly as enzyme sources (Shoji et al. 2007). Currently, approximately 90% of all industrial enzymes are produced via SmF using highly optimized and, in many cases, genetically engineered microbial strains (Driouch et al. 2010).

6.5.1 Primary Microbial Sources and Fermentation

The enzyme invertase is produced by the use of two main fermentation technologies, SmF and SSF. SmF refers to a traditional process where the cultivation of microorganisms takes place in a liquid medium. SSF comprises the development of microorganisms on some wet solid substrate usually with inert support such as polyurethane foam and it simulates their natural conditions. Studies have consistently demonstrated that SSF can give rise to much higher yields of invertase in comparison with SmF for certain fungal strains, e.g., *Aspergillus niger* (Guerrero-Urrutia et al. 2021). The greater productivity of SSF is not an unexpected result; it accrues directly from the physiological and biochemical contrasts between the two systems. In SSF, the microbial cells encounter lower water activity and an even more extreme state of nutrient limitation which approximates their normal condition on solid surfaces. This environmental stress can trigger a more robust and efficient secretion of hydrolytic enzymes, like invertase, to break down complex substrates. A key advantage of SSF is the reduced extent of catabolic repression. In SmF, high concentrations of glucose can repress the production of invertase, a common limitation in liquid cultures (Reddy and Maley 1990). This repression is much less pronounced in SSF, allowing for higher enzyme yields even in the presence of high sugar concentrations. Additionally, SSF can lead to earlier enzyme excretion, with peak production observed at 24 h compared to 36 h in SmF for a specific *A. niger* strain.

The vast majority of commercially produced invertase is of microbial origin, with key sources including yeasts and fungi. *Saccharomyces cerevisiae* is the preeminent strain for industrial production, largely because of its historical use in baking and fermentation, which has led to well-established, scalable cultivation techniques (Favaro et al. 2019). Other important microbial sources include the fungi such as *Aspergillus niger* and *Penicillium chrysogenum* and other yeasts such as *Candida utilis* and *Xanthophyllomyces dendrorhous* (Dworschack and Wickerham 1961; Linde et al. 2009). Several *C. utilis* strains were observed for invertase production using shake flask fermentation technique and strain KC-B10 showed the maximum invertase production. Research also observed that under ideal conditions extracellular and intracellular activities increased by 49% and 30%, respectively

(Qureshi and Aslam 2013). Linde et al. (2009) isolated and sequenced the *X. dendrorhous* β-fructofuranosidase gene (*Xd-INV*). The result of the study showed that it encodes a putative mature polypeptide which shares important uniqueness with other fungal, yeast, and plant β-fructofuranosidases (Linde et al. 2009).

The advent of recombinant DNA technology has further expanded production options. For instance, an invertase gene from *A. niger* can be expressed in a different, more robust microbial host like *Trichoderma reesei* to produce an enzyme preparation for food use, as detailed in recent FDA regulatory filings (Bergès et al. 1993). This approach leverages the best traits of different organisms: the desired enzyme from one source and the high productivity of a different host strain. Bacterial invertase production has been documented exclusively under SmF, with activities ranging from 0.35 IU/mL to 1399 IU/mL (Awad et al. 2013). Notably, *Lactobacillus brevis* Mm-6 achieved the highest reported bacterial invertase of 1399 IU/mL (Awad et al. 2013). Invertase yields vary widely across microbial taxa and fermentation strategies. Among fungi, the highest reported activity was 361.9 IU/mL from *Emericella nidulans* grown on rye flour under SmF conditions (Cuezzo de Ginés et al. 2000), and 144.39 IU/g from *Aspergillus niger* EM77 cultivated on sucrose under SSF (Kaur and Sharma 2005). Unusually, recombinant *A. niger* strains under SmF have achieved titers as high as 2800 IU/mL (Driouch et al. 2010). Additionally, *S. cerevisiae* NRRLY-12632 yielded 272.5 U/g under SSF with sucrose supplementation (Rashad and Nooman 2009).

6.5.2 Optimization Strategies for Enhanced Yield

Optimizing invertase production is critical for industrial viability. Traditional optimization methods, which involve changing one variable at a time, often fail to capture the complex interactions between different parameters. This limitation has led to the adoption of more advanced statistical techniques. Response surface methodology (RSM) is a powerful bioprocess engineering tool that uses statistical modeling to simultaneously analyze the effects of multiple variables and their interactions on enzyme yield. This approach, often employing a central composite design (CCD), significantly reduces the number of experimental trials needed to find optimal conditions. Plackett–Burman screening followed by a Box–Behnken response surface methodology is a significant approach for optimization of invertase production. Here, the regression model (R^2 = 97%) recognized sucrose (40 g/ml), yeast extract (5 g/ml), and peptone (4 g/ml) as optimal concentrations. Validation experiments confirmed these conditions, yielding 38.71 U/ml invertase activity a fourfold increase over the original medium demonstrating the efficiency of this statistical optimization approach (Ghasemi et al. 2014). Another study by Dinarvand et al. (2017) optimized the growth and production of inulinase (exo- and endo-) and invertase by *A. niger* ATCC 20611 using RSM with CCD. The model showed high accuracy ($R^2 > 0.90$), identifying pH and temperature as the most significant factors, with inoculum size and agitation speed having lesser effects. Optimal conditions (pH 6.5, 30 °C, 6% inoculum, 150 rpm) increased enzyme and biomass yields by

10–32 times compared to the one-factor-at-a-time method. This approach supports industrial-scale enzyme production (Dinarvand et al. 2017). By response surface methodology (RSM) also *A. niger* to optimize key physical parameters, identifying an optimal medium condition of pH 6.5, a temperature of 30 °C, an inoculum size of 6% (v/v), and an agitation speed of 150 rpm (Ire et al. 2018). These optimized conditions resulted in a remarkable 10- to 32-fold increase in enzyme production and biomass content compared to traditional one-factor-at-a-time methods. Genetic engineering is also another powerful approach to enhance invertase production. This may involve modification the host microorganism to create a high-yield strain. For example, the non-conventional yeast *Yarrowia lipolytica* which normally does not produce invertase can be made to express the SUC2 gene from *S. cerevisiae* under the control of a strong promoter. Hence in one case with prior construction yielding only 597 U/l extracellular invertase, this new strain was able to express over seven times as much extracellular invertase (4,519 U/l) as previously constructed strains (Lazar et al. 2013). Genetic improvement has also been achieved using parasexual crosses in *A. niger*. Invertase productivity values between 5 and 18 times higher than those recorded for SSF cultures have been obtained by researchers who dipped haploid mutants to form diploid strains (Montiel-González et al. 2002).

6.5.3 Purification and Stabilization Methods

After fermentation, a series of downstream processing steps are necessary to isolate and purify the enzyme. Since invertase is typically secreted into the periplasmic space in yeast, the initial step involves cell disruption using methods like sonication or high-pressure homogenization to release the enzyme (Kulshrestha et al. 2013). Purification of the crude enzyme extract can be achieved through traditional techniques, such as precipitation with ammonium sulfate, followed by column chromatography (e.g., DEAE-cellulose). Modern and more environmentally friendly alternatives, such as membrane filtration (microfiltration and ultrafiltration), have also proven highly effective. This method uses the selective permeability of membranes to separate the enzyme from other cellular components, leading to a significant increase in enzyme activity and specific activity. For example, one study using micro- and ultra-filtration increased enzyme activity from 153 IU/mL to 691 IU/mL (Durakli Velioglu et al. 2025).

For continuous industrial processes, the free enzyme's stability can be a limiting factor. To address this, invertase can be stabilized through immobilization, a process that fixes the enzyme to a solid support. A common and simple method is entrapment within a gel matrix, such as alginate beads. This strategy impressively improves the enzyme's thermal and storage stability. For instance, at 60 °C, the half-life of an immobilized invertase was 28 min, compared to just 8 min for the free enzyme. Immobilization also prevents the loss of activity during storage, with immobilized invertase retaining its activity for over 40 days at 2–4 °C in buffer. While immobilization can sometimes lead to a slightly reduced catalytic efficiency (higher Km and lower Vmax), the gains in stability, reusability, and ease of separation make it a highly cost-effective solution for industrial applications. Furthermore, the immobilization matrix can prevent unwanted side reactions, such as the formation of oligosaccharides that can occur with the free enzyme at high substrate concentrations (Husain et al. 1996).

6.6 Therapeutic Applications

The capacity of invertase to break the glycosidic bond between glucose and fructose makes it vital for the digestion of complex carbohydrates into glucose, a readily accessible rapid source of energy for numerous body processes (Kulshrestha et al. 2013). This natural enzyme is not only important in carbohydrate breakdown, but also in maintaining human health through helping fight disease, retarding the aging process, and inducing bodily rejuvenation. In the elderly, digestive efficiency tends to be impaired with decreased availability of this enzyme, which further reduces their ability to derive essential nutrients from foods. Apart from its role in digestion, invertase finds extensive use as a prebiotic in functional foods and the pharmaceutical sector (Toledo et al. 2019). In honey, invertase is a powerful antimicrobial substance with excellent antioxidant activity, catalyzing the conversion of glucose to hydrogen peroxide, which is involved in protection against bacterial infection and prevention of dangerous gut fermentation (Bugarova et al. 2021). Additionally, by generating pre-digested simple sugars, invertase minimizes stomach toxicity due to bacterial fermentation, thus assisting in the protection against ulcers and other gastrointestinal ailments. Research has indicated that naturally occurring invertase in honey is able to treat respiratory tract diseases like asthma, allergies, and bronchitis, and can also decrease the severity of the common cold and flu (Vitolo 2021). Enzymes present in honey, including invertase, can be effective for bone and stomach cancer patients (Kulshrestha et al. 2013). Certain fructosyl-transferase-containing invertases are industrially employed in the synthesis of fructooligosaccharides like 1-kestose, nystose, and fructofuranosyl nystose as non-carcinogenic sweeteners more soluble than sucrose (Canli et al. 2011). Besides, invertase has been reported to decrease blood cholesterol and ammonia levels as well as to enhance the proliferation of health-imparting *Bifidobacterium* in the colon of human beings (Jorge and Guimarães 2013; Laue et al. 2019).

6.7 Conclusion

Microbial sacrosidase has changed over time from a simple historical biochemical topic to a sophisticated, well-developed industrial biocatalyst. Its function of hydrolyzing sucrose into invert sugar remains critical in the food and beverage sector as it helps produce soft-centered candies and fondants which are non-crystallizable and hygroscopic. Its production has transitioned from traditional submerged fermentation to more efficient solid-state fermentation and the application of rigorous statistical and genetic engineering tools. Modern techniques like membrane filtration and stabilization through immobilization have dramatically improved the enzyme's purification, reusability, and operational stability. The most visible change in invertase is seen in its newer uses. Besides the traditional hydrolysis function, invertase can now also act as a fructosyltransferase for the production of prebiotic FOS, extending the scope of functional foods and medicines. This allows for the precise modification of its properties, such as pH and temperature optima, to suit a wide range of specific industrial demands. The regulatory landscape, as shown by recent FDA GRAS notices, is also adapting to these advancements, providing a clear pathway for innovative, safe, and effective enzyme preparations. Invertase's journey from a historical curiosity to a modern, versatile biocatalyst reflects the dynamic interplay between fundamental biochemical knowledge and cutting-edge biotechnological innovation, ensuring its enduring importance in the bioeconomy.

References

Aleksanyan ER, Markosyan LS (1986) Microbial -fructofuranosidase (survey). Appl Biochem Microbiol 22:133–145

Álvaro-Benito M, Abreu M, Fernandez-Arrojo L, Plou FJ, Jimenez-Barbero J, Ballesteros A, Polaina J, Fernandez-Lobato M (2007) Characterization of a β-fructofuranosidase from *Schwanniomyces occidentalis* with transfructosylating activity yielding the prebiotic 6-kestose. J Biotechnol 132(1):75–81

Andjelković U, Pićuric S, Vujčic Z (2010) Purification and characterization of *Saccharomyces cerevisiae* external invertase isoforms. Food Chem 120:799–804

Antosova M, Polakovic M (2001) Fructosyltransferases: the enzymes catalyzing production of fructooligosaccharides. Chem Pap 55(6):350–358

Araújo VPB, Araújo TK, Sousa KMN, Albuquerque WWC, Nascimento AKCD, Cardoso KBB, Nascimento TP, Batista JMDS, Cavalcanti MTH, Porto ALF, Costa RMPB (2023) A novel β-fructofuranosidase produced by *Penicillium citreonigrum* URM 4459: purification and biochemical features. Prep Biochem Biotechnol 53(8):906–913. https://doi.org/10.1080/10826068.2022.2158472

Ávila-Fernández Á, Cuevas-Juárez E, Rodríguez-Alegría ME, Olvera C, López-Munguía A (2016) Functional characterization of a novel β-fructofuranosidase from *Bifidobacterium longum* subsp. infantis ATCC 15697 on structurally diverse fructans. *J Appl Microbiol* 121(1):263–276. https://doi.org/10.1111/jam.13154

Awad GEA, Hassan A, Eman WEG, Wafaa AH, Mona AE, Magdy MME (2013) Production optimization of invertase by *lactobacillus brevis* Mm-6 and its immobilization on alginate beads. Carbohydr Polym 93:740–746

Barbosa PMG, de Morais TP, de Andrade Silva CA, da Silva Santos FR, Garcia NFL, Fonseca GG, Ribeiro Liete RS, da Paz MF (2018) Biochemical characterization and evaluation of invertases produced from *Saccharomyces cerevisiae* CAT-1 and *Rhodotorula mucilaginosa* for the production of fructooligosaccharides. Prep Biochem Biotechnol 48(6):506–513. https://doi.org/10.1080/10826068.2018.1466155

Batista RD, Melo FG, do Amaral Santos CCA, de Paula-Elias FC, Perna RF, Xavier MCA, Villalba Morales SA, de Almeida AF (2021) Optimization of β-Fructofuranosidase production from Agrowaste by *Aspergillus carbonarius* and its application in the production of inverted sugar. Food Technol Biotechnol 59(3):306–313. https://doi.org/10.17113/ftb.59.03.21.6934

Belcarz A, Ginalska G, Lobarzewski J, Penel C (2002) The novel non-glycosylated invertase from *Candida utilis* (the properties and the conditions of production and purification). Biochim Biophys Acta 1594(1):40–53. https://doi.org/10.1016/s0167-4838(01)00279-5

Belmonte-Izquierdo Y, Salomé-Abarca LF, González-Hernández JC, López MG (2023) Fructooligosaccharides (FOS) production by microorganisms with Fructosyltransferase activity. Fermentation 9:968. https://doi.org/10.3390/fermentation9110968

Bergès T, Barreau C, Peberdy JF, Boddy LM (1993) Cloning of an *Aspergillus Niger* invertase gene by expression in *Trichoderma reesei*. Curr Genet 24(1):53–59

Bhatti HN, Asgher M, Abbas A, Nawaz R, Sheikh MA (2006) Studies on kinetics and Thermostability of a novel acid Invertase from *Fusarium solani*. J Agric Food Chem 54(13):4617–4623

Boer E, Wartmann T, Luther B, Manteuffel R, Bode R, Gellissen G, Kunze G (2004) Characterization of the AINV gene and the encoded invertase from the dimorphic yeast *Arxula adeninivorans*. Antonie Van Leeuwenhoek 86(2):121–134. https://doi.org/10.1023/B:ANTO.0000036135.69810.df

Buchner E (1897) Alcoholic fermentation without yeast cells. Ber Dtsch Chem Ges 30(1):117–124

Bugarova V, Godocikova J, Bucekova M, Brodschneider R, Majtan J (2021) Effects of the carbohydrate sources nectar, sucrose and invert sugar on antibacterial activity of honey and bee-processed syrups. Antibiotics 10(8):985. https://doi.org/10.3390/antibiotics10080985

Canli O, Erdal S, Taskin M, Kurbanoglu EB (2011) Effects of extremely low magnetic field on the production of invertase by *Rhodotorula glutinis*. Toxicol Ind Health 27(1):35–39

Chen JS, Saxton J, Hemming FW, Peberdy JF (1996) Purification and partial characterization of the high and low molecular weight form (S- and F-form) of invertase secreted by *Aspergillus nidulans*. Biochim Biophys Acta 1296:207–218

Chen C, Zhou F, Ren J, Ai L, Dong Y, Wu Z, Liu Z, Chen W, Guo B (2014) Cloning, expression and functional validation of a β-fructofuranosidase from *Lactobacillus plantarum*. Process Biochem 49(5):758–767. https://doi.org/10.1016/j.procbio.2014.02.013

Chen Z, Shen Y, Xu J (2023) A strategy for rapid acquisition of the β-D-Fructofuranosidase gene through chemical synthesis and new function of its encoded enzyme to improve gel properties during yogurt processing. Foods 12(8):1704. https://doi.org/10.3390/foods12081704

Cuervo-Fernandez R, Ottoni CA, Silva ES, Matsubara RMS, Carter JM, Magossi LR, Wada MA, Andrade MF, Maresma BG, Maiorano AE (2007) Screening of β-fructofuranosidase-producing microorganisms and effect of pH and temperature on enzymatic rate. Appl Microbiol Biotechnol 74:87–93. https://doi.org/10.1007/s00253-006-0803-x

Cuezzo de Ginés S, Maldonado M, Font de Valdez G (2000) Purification and characterization of invertase from *Lactobacillus reuteri* CRL 1100. Curr Microbiol 40(3):181–184. https://doi.org/10.1007/s002849910036

Dinarvand M, Rezaee M, Foroughi M (2017) Optimizing culture conditions for production of intra and extracellular inulinase and invertase from *Aspergillus Niger* ATCC 20611 by response surface methodology (RSM). Braz J Microbiol 48(3):427–441

Driouch H, Roth A, Dersch P, Wittmann C (2010) Optimized bioprocess for production of fructofuranosidase by recombinant *Aspergillus Niger*. Appl Microbiol Biotechnol 87(6):2011–2024

Du LQ, Pang H, Hu YY, Wei YT, Huang RB (2010) Expression and characterization in *E. coli* of a neutral invertase from a metagenomic library. World J Microbiol Biotechnol 26:419–428. https://doi.org/10.1007/s11274-009-0184-2

Du L, Pang H, Wang Z, Lu J, Wei Y, Huang R (2013) Characterization of an invertase with pH tolerance and truncation of its N-terminal to shift optimum activity toward neutral pH. PLoS One 8(4):e62306. https://doi.org/10.1371/journal.pone.0062306

Durakli Velioglu S, Bagci U, Guner KG, Gulec HA, Velioglu HM (2025) Green purification of invertase from ultrasonicated sifted baker's yeast by membrane filtration: a comparative study. Molecules 30(12):2663. https://doi.org/10.3390/molecules30122663

Dworschack RG, Wickerham LJ (1961) Production of extracellular and total invertase by *Candida utilis*, *Saccharomyces cerevisiae*, and other yeasts. Appl Microbiol 9(4):291–294

Farine S, Versluis C, Bonnici PJ, Heck A, L'homme C, Puigserver A, Biagini A (2001) Application of high-performance anion exchange chromatography to study invertase-catalysed hydrolysis of sucrose and formation of intermediate fructan products. Appl Microbiol Biotechnol 55(1):55–60. https://doi.org/10.1007/s002530000493

Favaro L, Jansen T, van Zyl WH (2019) Exploring industrial and natural *Saccharomyces cerevisiae* strains for the bio-based economy from biomass: the case of bioethanol. Crit Rev Biotechnol 39(6):800–816

Fernandes P (2025) Fructooligosaccharides (FOSs): a condensed overview. Compounds 5(2):8. https://doi.org/10.3390/compounds5020008

Fernandes ML, Jorge JA, Guimarães LH (2018) Characterization of an extracellular β-D-fructofuranosidase produced during solid-state fermentation (SSF) of cassava husk. J Food Biochem 42:e12443

Fernandez RC, Maresma BG, Juarez A, Martínez J (2004) Production of fructooligosaccharides by β-fructofuranosidase from *Aspergillus* sp 27H. J Chem Technol Biotechnol. https://doi.org/10.1002/JCTB.967. Corpus ID: 260891294

FSANZ (2021) Schedule 18. Processing aids (F2021C00092). In: Australia New Zealand food standards code. Australian Government, Canberra. https://www.legislation.gov.au/Details/F2021C00092. Accessed 11 June 2021

Ghasemi Y, Mohkam M, Ghasemian A, Rasoul-Amini S (2014) Experimental design of medium optimization for invertase production by *Pichia* sp. J Food Sci Technol 51(2):267–275

Gimeno-Pérez M, Linde D, Fernández-Arrojo L, Plou FJ, Fernández-Lobato M (2015) Heterologous overproduction of β-fructofuranosidase from yeast *Xanthophyllomyces dendrorhous*, an enzyme producing prebiotic sugars. Appl Microbiol Biotechnol 99(8):3459–3467. https://doi.org/10.1007/s00253-014-6145-1

Gimeno-Pérez M, Merdzo Z, Castillo-Rosa E, Hijas CM, Fernández-Lobato M (2021) The β-fructofuranosidase from *Rhodotorula dairenensis*: molecular cloning, heterologous expression, and evaluation of its transferase activity. Catalysts 11(4):476. https://doi.org/10.3390/catal11040476

Gonçalves HB, Jorge JA, Oliveira WP, Souza CRF, Guimarães LHS (2013) Extracellular β-fructofuranosidase from *Fusarium graminearum*: stability of the spray-dried enzyme in the presence of different carbohydrates. J Microencapsul 30(7):624–631. https://doi.org/10.3109/02652048.2013.770097

Goud KG, Chaitanya K, Reddy G (2013) Enhanced production of β-D-fructofuranosidase by *Saccharomyces cerevisiae* using agro-industrial wastes as substrates. Biocatal Agric Biotechnol 2(4):385–392. https://doi.org/10.1016/j.bcab.2013.08.001

GrootWassink JWD, Fleming SE (1980) Non-specific β-fructofuranosidase (inulase) from *Kluyveromyces fragilis*: batch and continous fermentation, simple recovery method and some industrial properties. Enzym Microb Technol 2(1):45–53. https://doi.org/10.1016/0141-0229(80)90007-1

Guerrero-Urrutia C, Volke-Sepulveda T, Figueroa-Martinez F, Favela-Torres E (2021) Solid-state fermentation enhances inulinase and invertase production by *Aspergillus brasiliensis*. Process Biochem 108:169–175

Guimaraes LHS, Terenzi HF, Polizeli MLTM, Jorge JA (2007) Production and characterization of thermostable extracellular β-D-fructofuranosidase produced by *Aspergillus ochraceus* with agroindustrial residues as carbon sources. Enyzme Microb Technol 42:52–57

Gutiérrez-Alonso P, Fernández-Arrojo L, Plou FJ, Fernández-Lobato M (2009) Biochemical characterization of a β-fructofuranosidase from *Rhodotorula dairenensis* with transfructosylating activity. FEMS Yeast Res 9:768–773

Han Z, Kong Q, Xu J (2025) Multi-target antifungal action of engineered binuclear zinc cluster peptides against *Aspergillus flavus*: integrating membrane disruption, transcriptional interference, and metabolic dysregulation. Food Res Int 221:117254

Health Canada (2021) List of permitted food enzymes. Lists of Permitted Food Additives, No. 5. Health Canada. https://www.canada.ca/en/health-canada/services/food-nutrition/food-safety/food-additives/lists-permitted/5-enzymes.html. Accessed 11 June 2021

Henrissat B, Bairoch A (1996) Updating the sequence-based classification of glycosyl hydrolases. Biochem J 316:695–699

Hirayama M, Sumi N, Hidaka H (1989) Purification and properties of fructo-oligosaccharides producing beta-fructofuranosidase from *Aspergillus Niger* ATCC 20611. Agric Biol Chem 53:667–673

Husain S, Jafri F, Saleemuddin M (1996) Effects of chemical modification on the stability of invertase before and after immobilization. Enzym Microb Technol 18(4):275–280

Ire FS, Aguguo VJ, Ezebuiro V (2018) Optimization of invertase from *Aspergillus Niger* grown on low-cost agricultural wastes by response surface methodology (RSM). J Adv Microbiol 13(4):1–15

Janer C, Rohr LM, Peláez C, Laloi M, Cleusix V, Requena T, Meile L (2004) Hydrolysis of oligofructoses by the recombinant β-fructofuranosidase from *Bifidobacterium lactis*. Syst Appl Microbiol 27(3):279–285

Jedrzejczak-Krzepkowska M, Tkaczuk KL, Bielecki S (2011) Biosynthesis, purification and characterization of β-fructofuranosidase from *Bifidobacterium longum* KN29.1. Process Biochem 46(10):1963–1972. https://doi.org/10.1016/j.procbio.2011.07.005

Jorge JA, Guimarães LHS (2013) Production of invertases by anamorphic (*Aspergillus nidulans*) and teleomorphic (*Emericela nidulans*) fungi under submerged fermentation using rye flour as carbon source. Adv Microbiol 3(5):421–429

Kaur N, Sharma AD (2005) Production, optimization and characterization of extracellular invertase by an actinomycete strain. J Sci Ind Res 64(7):515–519

Kulshrestha S, Tyagi P, Sindhi V, Yadavilli KS (2013) Invertase and its applications–a brief review. J Pharm Res 7(9):792–797

Kurakake M, Ogawa K, Sugie M, Takemura A, Sugiura K, Komaki T (2008) Two types of b-fructofuranosidases from *Aspergillus oryzae* KB. J Agric Food Chem 56:591–596

Kurimoto M, Tsusaki K, Kubota M, Fukuda S, Tsujisaka Y (1999) Cloning and sequencing of the beta-fructofuranosidase gene from *Bacillus* sp. V230. Biosci Biotechnol Biochem 63(6):1107–1111. https://doi.org/10.1271/bbb.63.1107

Laue C, Ballance S, Knutsen SH, Papazova E, Soeth E, Pannenbeckers A, Schrezenmeir J (2019) Glycemic response to low sugar apple juice treated with invertase, glucose oxidase and catalase. Eur J Clin Nutr 73(10):1382–1391

Lazar Z, Rossignol T, Verbeke J, Crutz-Le Coq AM, Nicaud JM, Robak M (2013) Optimized invertase expression and secretion cassette for improving *Yarrowia lipolytica* growth on sucrose for industrial applications. J Ind Microbiol Biotechnol 40(11):1273–1283

Legifrance (2006) Arrêté du 19 octobre 2006 relatif à l'emploi d'auxiliaires technologiques dans la fabrication de certaines denrées alimentaires. NOR: ECOC0600115A [Article Annexe I C, enzymes]. J Off Repub Fr. http://www.legifrance.gouv.fr/affichTexte.do;jsessionid=C9CD6B3BA59CBF9B2355B79A66A0FBFE.tpdjo04v_3?idSectionTA=LEGISCTA000020667473&cidTexte=JORFTEXT000000271061&dateTexte=20140314. Accessed 11 June 2021

Lincoln L, More SS (2017) Bacterial invertases: occurrence, production, biochemical characterization, and significance of transfructosylation. J Basic Microbiol 57(10):803–813. https://doi.org/10.1002/jobm.201700269

Lincoln L, More SS (2018) Comparative evaluation of extracellular β-d-fructofuranosidase in submerged and solid-state fermentation produced by newly identified Bacillus subtilis strain. J Appl Microbiol 125(2):441–456. https://doi.org/10.1111/jam.13881

Linde D, Macias I, Fernández-Arrojo L, Plou FJ, Jiménez A, Fernández-Lobato M (2009) Molecular and biochemical characterization of a β-fructofuranosidase from *Xanthophyllomyces dendrorhous*. Appl Environ Microbiol 75(4):1065–1073. https://doi.org/10.1128/AEM.02061-08

Liu J, Cheng Y, Ruan M, Ye Q, Wang R, Yao Z, Zhou G, Li Z, Liu C, Wan H (2025) Roles and regulations of acid Invertases in plants: current knowledge and future perspectives. Plants 14(3):320. https://doi.org/10.3390/plants14030320

Lombard V, Ramulu HG, Drula E, Coutinho PM, Henrissat B (2013) The carbohydrate-active enzymes database (CAZy) in 2013. Nucleic Acids Res 42:490–495

Lorentz K, Lienhard S, Sturm A (1995) Structural organization and differential expression of carrot β-fructofuranosidase genes: identification of a gene coding for a flower bud-specific isoenzyme. Plant Mol Biol 28:189–194

Lorenzoni AS, Aydos LF, Klein MP, Rodrigues RC, Hertz PF (2014) Fructooligosaccharides synthesis by highly stable immobilized β-fructofuranosidase from *Aspergillus aculeatus*. Carbohydr Polym 103:193–197. https://doi.org/10.1016/j.carbpol.2013.12.038

Manoochehri H, Hosseini NF, Saidijam M, Taheri M, Rezaee H, Nouri F (2020) A review on invertase: its potentials and applications. Biocatal Agric Biotechnol 25:101599. https://doi.org/10.1016/j.bcab.2020.101599

Maresma BG, Castillo BG, Fernández RC, Silva ESD, Maiorano AE, Rodrigues MFDA (2010) Mutagenesis of *Aspergillus oryzae* IPT-301 to improve the production of β-fructofuranosidase. Braz J Microbiol 41:186–195. https://doi.org/10.1590/S1517-83822010000100027

Matei GM, Matei S, Pele M, Dumitrescu F, Matei A (2017) Invertase production by fungi, characterization of enzyme activity and kinetic parameters. Rev Chim 68(10):2205–2208. https://doi.org/10.37358/RC.17.10.5856

MFDS (2019) [II. Food additives and mixed preparations 4. Component specification. A. Food Additives. Invertase]. In: Korea food additives code. (All designated chemicals and some natural additives are currently regulated by Food Additives Code. It includes specifications, standards and general test methods for each additives). Ministry of Food and Drug Safety. https://www.foodsafetykorea.go.kr/foodcode/04_03.jsp?idx=48. Accessed 11 June 2021

Montiel-González AM, Fernández FJ, Viniegra-González G, Loera O (2002) Invertase production on solid-state fermentation by *Aspergillus Niger* strains improved by parasexual recombination. Appl Biochem Biotechnol 102(1):63–70

Moreira LA, Cavalcanti Oliveira AH, Guimarães LHS (2024) Production and characterization of an extracellular Mn2+ activated β-D-fructofuranosidase from *Aspergillus labruscus* ITAL 28.255 with transfructosylating activity. J Appl Biol Biotechnol 12(3):236–243. https://doi.org/10.7324/JABB.2024.160066

Moreno S, Sanchez Y, Rodriguez L (1990) Purification and characterization of the invertase from *Schizosaccharomyces pombe*. A comparative analysis with the invertase from *Saccharomyces cerevisiae*. Biochem J 267:697–702

Muramatsu K, Onodera S, Kikuchi M, Shiomi N (1992) The production of βFructofuranosidase from *Bifidobacterium* spp. Biosci Biotech Bioch 56(9):1451–1454. https://doi.org/10.1271/bbb.56.1451

Mussatto SI, Aguilar CN, Rodrigues LR, Teixeira JA (2009) Fructooligosaccharides and β-fructofuranosidase production by *Aspergillus japonicus* immobilized on lignocellulosic materials. J Mol Catal B Enzym 59(1–3):76–81. https://doi.org/10.1016/j.molcatb.2009.01.005

Nadeem H, Rashid MH, Siddique MH, Azeem F, Muzammil S, Javed MR, Ali MA, Rasul I, Riaz M (2015) Microbial invertases: a review on kinetics, thermodynamics, physiochemical properties. Process Biochem 50(8):1202–1210

Nascimento AKC, Nobre C, Cavalcanti MTH, Teixeira JA, Porto ALF (2016) Screening of fungi from the genus *Penicillium* for production of β- fructofuranosidase and enzymatic synthesis of fructooligosaccharides. J Mol Catal B Enzym 134(A):70–78. https://doi.org/10.1016/j.molcatb.2016.09.005

Naumoff DG (2001) β-Fructosidase superfamily: homology with some α-l-arabinases and β-D-xylanases. Protein 42:66–76

Nguyen QD, Rezessy-Szabo JM, Bhat MK, Hoschke A (2005) Purification and some properties of β-fructofuranosidase from *Aspergillus Niger* IMI303386. Process Biochem 40:2461–2466

Niu L, Liu L, Wang W (2020) Digging for stress-responsive cell wall proteins for developing stress-resistant maize. Front Plant Sci 11:576385

O'Mullan PJ, Chase T, Eveleigh DE (1992) Purification and some properties of extracellular invertase B from *Zymomonas mo*bilis. Appl Microbiol Biotechnol 38:341–346

Oliveira RL, Silva MF, Converti A, Porto TS (2020) Production of β fructofuranosidase with transfructosylating activity by *Aspergillus tamarii* URM4634 solid-state fermentation on agroindustrial byproducts. Int J Biol Macromol 144:343–350

Onitilo VO, Sanusi JF, Adebayo-Tayo BC (2022) Production, purification and characterization of B-D-Fructofuranosidase from *Fusarium oxysporum* isolated from grape juice and soil sample from sugar cane dumpsite. Adv Multidiscip Sci Res 8(3):1–12. https://doi.org/10.22624/AIMS/V8N3P1

Osiebe O, Adewale IO, Omafuvbe BO (2023) Production and characterization of intracellular invertase from *Saccharomyces cerevisiae* (OL629078.1), using cassava-soybean as a cost-effective substrate. Sci Rep 13:16295. https://doi.org/10.1038/s41598-023-43502-2

Ottoni CA, Cuervo-Fernández R, Piccoli RM, Moreira R, Guilarte-Maresma B, Sabino da Silva E, Rodrigues MFA, Maiorano AE (2012) Media optimization for β-fructofuranosidase production by *Aspergillus oryzae*. Braz J Chem Eng 29:49–59. https://doi.org/10.1590/S0104-66322012000100006

Ozimek LK, Kralj S, van der Maarel M, Dijkhuizen L (2006) The levansucrase and inulosucrase enzymes of *Lactobacillus reuteri* 121 catalyse processive and non-processive transglycosylation reactions. Microbiol-Sgm 152:1187–1196

Pérez JA, Rodríguez J, Rodríguez L, Ruiz T (1996) Cloning and sequence analysis of the invertase gene INV 1 from the yeast *Pichia anomala*. Curr Genet 29(3):234–240. https://doi.org/10.1007/BF02221553

Plascencia-Espinosa M, Santiago-Hernández A, Pavón-Orozco P, Becerra V, Trejo-Estrada VS, Sosa-Peinado A, Benitez-Cardoza CG, Hidalgo-Lara ME (2014) Effect of deglycosylation on the properties of thermophilic invertase purified from the yeast *Candida guilliermondii* MpIIIa. Process Biochem 49(9):1480–1487. https://doi.org/10.1016/j.procbio.2014.05.022

Punt PJ, Biezen van N, Conesa A, Albers A, Mangnus J, Hondel van den C (2002) Filamentous fungi as cell factories for heterologous protein production. Trends Biotechnol 20(5):200–206. https://doi.org/10.1016/S0167-7799(02)01933-9

Qureshi FZ, Aslam A (2013) A comparative study of nutritional and environmental factors affecting extracellular and intracellular invertase production in *Candida utilis*. Pak J Bot 45(2):681–686

Ramírez-Escudero M, Gimeno-Pérez M, González B, Linde D, Merdzo Z, Fernández-Lobato M, Sanz-Aparicio J (2016) Structural analysis of β-Fructofuranosidase from *Xanthophyllomyces dendrorhous* reveals unique features and the crucial role of N-glycosylation in Oligomerization and activity. J Biol Chem 291(13):6843–6857. https://doi.org/10.1074/jbc.M115.708495

Rashad MM, Nooman MU (2009) Production, purification and characterization of extracellular invertase from *Saccharomyces cerevisiae* NRRL Y-12632 by solid-state fermentation of red carrot residue. Aust J Basic Appl Sci 3(3):1910–1919

Rashad MM, Mahmoud AEE, Desouky MA, Nooman MU (2006) Purification and characterization of extra and intracellular â-fructofuranosidase from Saccharomyces cerevisiae growing on *Eichhornia crassipes* leaf extract. Dtsch Lebensmitt Rundsch 102:157–166

Reddy VA, Maley F (1990) Identification of an active-site residue in yeast invertase by affinity labeling and site-directed mutagenesis. J Biol Chem 265(19):10817–10820

Rodrigo-Frutos D, Jiménez-Ortega E, Piedrabuena D, Ramírez-Escudero M, Míguez N, Plou FJ, Sanz-Aparicio J, Fernández-Lobato M (2021) New insights into the molecular mechanism behind mannitol and erythritol fructosylation by β-fructofuranosidase from *Schwanniomyces occidentalis*. Sci Rep 11:7158. https://doi.org/10.1038/s41598-021-86568-6

Rodriguez J, Perez JA, Ruiz T, Rodriguez L (1995) Characterization of the invertase from Pichia anomala. Biochem J 306:235–239. https://doi.org/10.1042/bj3060235

Rubio MC, Maldonado MC (1995) Purification and characterisation of invertase from *aspergillus Niger*. Curr Microbiol 31:80–83

Rubio MC, Runco R, Navarro AR (2002) Invertase from a strain of *Rhodotorula glutinis*. Phytochemistry 61(6):605–609. https://doi.org/10.1016/S0031-9422(02)00336-9

Ryan SM, Fitzgerald GF, van Sinderen D (2005) Transcriptional regulation and characterization of a novel β-Fructofuranosidase-encoding gene from *Bifidobacterium breve* UCC2003. Appl Environ Microbiol 71:3475–3482. https://doi.org/10.1128/AEM.71.7.3475-3482.2005

Salim RG, Abo-Sereh NA, Khalil BE (2020) Optimization and molecular characterization of novel Aspergillus spp. producing invertase enzyme. *Egypt Pharm J* 19(4):321–329. https://doi.org/10.4103/epj.epj_34_20

Sangeetha PT, Ramesh MN, Prapulla SG (2005) Maximization of fructooligosaccharide production by two stage continuous process and its scale up. J Food Eng 68(1):57–64. https://doi.org/10.1016/j.jfoodeng.2004.05.022

Schallmey M, Singh A, Ward OP (2004) Developments in the use of a species for industrial production. Can J Microbiol 50:1–7

Shankar T, Thangamathi P, Rama R, Sivakumar T (2013) Optimization of invertase production using *Saccharomyces cerevisiae* MK under varying cultural conditions. Int J Biochem Biophys 1(3):47–56. https://doi.org/10.13189/ijbb.2013.010301

Shiomi N (1992) Content of carbohydrate and activities of fructosyltransferase and invertase in asparagus roots during the fructo-oligosaccharide- and fructo-polysaccharide accumulating season. New Phytol 122(3):421–432. https://doi.org/10.1111/j.1469-8137.1992.tb00069.x

Shoji H, Sugimoto T, Hosoi K, Shibata K, Tanabe M, Kawatsura K (2007) Simultaneous production of glucoamylase and acid-stable α-amylase using novel submerged culture of *Aspergillus kawachii* NBRC4308. J Biosci Bioeng 103(2):203–205

Siddiqa A, Amin G (2022) Invertase enzyme. Int J Adv Biochem Res 6(2):120–126. https://doi.org/10.33545/26174693.2022.v6.i2b.148

Smaali I, Jazzar S, Soussi A, Muzard M, Aubry N, Marzouki MN (2012) Enzymatic synthesis of fructooligosaccharides from date by-products using an immobilized crude enzyme preparation of β-D-fructofuranosidase from *Aspergillus awamori* NBRC 4033. Biotechnol Bioprocess Eng 17:385–392. https://doi.org/10.1007/s12257-011-0388-9

Steensels J, Verstrepen KJ (2014) Taming wild yeast: potential of conventional and nonconventional yeasts in industrial fermentations. Ann Rev Microbiol 68:61–80. https://doi.org/10.1146/annurev-micro-091213-113025

Sturm A, Tang GQ (1999) The sucrose-cleaving enzymes of plants are crucial for development, growth and carbon partitioning. Trends Plant Sci 4:401–407

Sturm A, Hess D, Lee HS, Lienhard S (1999) Neutral invertase is a novel type of sucrose-cleaving enzyme. Physiol Plant 107:159–165. https://doi.org/10.1034/j.1399-3054.1999.100202.x

Tanaka T, Matsumoto S, Yamada M, Yamada R, Matsuda FK (2013) A display of active beta-glucosidase on the surface of *Schizosaccharomyces pombe* cells using novel anchor proteins. Appl Microbiol Biotechnol 97:4343–4352. https://doi.org/10.1007/s00253-013-4733-0

Timerman AP (2012) The isolation of invertase from baker's yeast–an introduction to protein purification strategies. InTech, Wisconsin

Tódero LM, Rechia CGV, Guimarães LHS (2019) Production of short-chain fructooligosaccharides (scFOS) using extracellular β-D-fructofuranosidase produced by *Aspergillus thermomutatus*. J Food Biochem 43:e12937. https://doi.org/10.1111/jfbc.12937

Toledo LET, García DM, Cruz EP, Intriago LMR, Pérez JN, Chanfrau JMP (2019) Fructosyltransferases and invertases: useful enzymes in the food and feed industries. In: Enzymes in food biotechnology, pp 451–469

Treem WR (2012) Clinical aspects and treatment of congenital sucrase-isomaltase deficiency. J Pediatr Gastroenterol Nutr 55(S2):S7–S13

Treem WR, McAdams L, Stanford L, Kastoff G, Justinich C, Hyams J (1999) Sacrosidase therapy for congenital sucrase-isomaltase deficiency. J Pediatr Gastroenterol Nutr 28(2):137–142

Trollope KM, Görgens JF, Volschenk H (2015) Semirational directed evolution of loop regions in *Aspergillus japonicus* β-fructofuranosidase for improved fructooligosaccharide production. Appl Environ Microbiol 81:7319–7329. https://doi.org/10.1128/AEM.02134-15

U.S. FDA (2018) Enzyme preparations used in food (partial list). Center for Food Safety and Applied Nutrition (CFSAN). https://www.fda.gov/food/generally-recognized-safe-gras/enzyme-preparations-used-food-partial-list. Accessed 16 June 2021

Van den Ende W, Van Laere A (1996) Fructan synthesizing and degrading activities in chicory roots (*Cichorium intybus* L.) during growth, storage and forcing. J Plant Physiol 149(1–2):43–50. https://doi.org/10.1016/S0176-1617(96)80171-4

Veana F, Fuentes-Garibay JA, Aguilar CN, Rodríguez-Herrera R, Guerrero-Olazarán M, Viader-Salvadó JM (2014) Gene encoding a novel invertase from a xerophilic *Aspergillus Niger* strain and production of the enzyme in *Pichia pastoris*. Enzym Microb Technol 63:28–33. https://doi.org/10.1016/j.enzmictec.2014.05.001

Veana F, Flores-Gallegos AC, Gonzalez-Montemayor AM, Michel-Michel M, Lopez-Lopez L, Aguilar-Zarate P, Ascacio-Valdés JA, Rodríguez-Herrera R (2018) Invertase: an enzyme with importance in confectionery food industry. In: Enzymes in food technology: improvements and innovations. Springer, Singapore, pp 187–212

Vitolo M (2021) Overview on invertase. World J Pharm Pharm Sci 10:49–73

Vo TD, Meetro J, Lynch B, Tafazoli S, Ichihara A, Chikamatsu G (2021) Safety assessment of β-fructofuranosidase from *Aspergillus brunneoviolaceus*. Toxicol Res Appl 5. https://doi.org/10.1177/23978473211055361

Warchol M, Perrin S, Grill JP, Schneider F (2002) Characterization of a purified beta-fructofuranosidase from *Bifidobacterium infantis* ATCC 15697. Lett Appl Microbiol 35(6):462–467. https://doi.org/10.1046/j.1472-765x.2002.01224.x

Win TT, Isono N, Kusnadi Y, Watanabe K, Obae K, Ito H, Matsui H (2004) Enzymatic synthesis of two novel non-reducing oligosaccharides using transfructosilation activity with beta-fructofuranosidase from *Arthrobacter globiformis*. Biotechnol Lett 26:499–503. https://doi.org/10.1023/b:bile.0000019557.44196.63

Wolska-Mitaszko B, Jaroszuk-Scisei J, Pszeniczna K (2007) Isoforms of trehalase and invertase of *Fusarium oxysporum*. Mycol Res III:456–465

Xu Q, Zheng X, Huang M, Wu M, Yan Y, Pan J, Yang Q, Duan CJ, Liu JL, Feng JX (2015) Purification and biochemical characterization of a novel β-fructofuranosidase from *Penicillium oxalicum* with transfructosylating activity producing neokestose. Process Biochem 50(8):1237–1246. https://doi.org/10.1016/j.procbio.2015.04.020

Yanase H, Fukushi H, Ueda N, Maeda Y, Toyoda A, Tonomura K (1991) Cloning, sequencing, and characterization of the intracellular invertase gene from *Zymomonas mobilis*. Agric Biol Chem 55:1383–1390

Yoon MH, Choi WY, Kwon SJ, Yi SH, Lee DH, Lee JS (2007) Purification and properties of intracellular invertase from alkalophilic and thermophilic *Bacillus cereus* TA-11. J Appl Biol Chem 50:196–201

Yoshikawa J, Amachi S, Shinoyama H, Fujii T (2006) Multiple β-fructofuranosidases by *Aureobasidium pullulans* DSM2404 and their roles in fructooligosaccharide production. FEMS Microbiol Lett 265:159–163. https://doi.org/10.1111/j.1574-6968.2006.00488.x

Zhou J, He L, Gao Y, Han N, Zhang R, Wu Q, Li J, Tang X, Xu B, Ding J, Huang Z (2016) Characterization of a novel low-temperature-active, alkaline and sucrose-tolerant invertase. Sci Rep 6:32081. https://doi.org/10.1038/srep32081

Zorn H, Barat JM, Bolognesi C, Catania F, Gadermaier G, Greiner R, Mayo B, Mortensen A et al (2024) Safety evaluation of the food enzyme β-fructofuranosidase from the non-genetically modified *Saccharomyces cerevisiae* strain CNCM I-3399. EFSA J 22(12):e9128. https://doi.org/10.2903/j.efsa.2024.9128

Harnessing the Potential of Microbial Lactase for Treating Lactose Intolerance

7

Richa Salwan, Rhydum Sharma, Randhir Kaur, Vivek Sharma, and Ghazala Mustafa

Abstract

Lactose intolerance is a widespread digestive disorder caused by the deficiency of lactase (β-galactosidase) and responsible for the catabolism of dairy products. This chapter provides a comprehensive overview of key advancements and explores significant microbial lactases as powerful therapeutic solutions due to their advantages such as high yields, cost-effective fermentation-based production, and ease of manipulation. Emphasis has been given on the factor responsible for enzyme stability through immobilization, microencapsulation, and the development of acid- and bile-resistant formulations. The chapter also explores the recent developments to improve the catalytic efficiency and functional properties of microbial lactases. Attempt has been made on the safety assessment, regulatory approval, and the future prospects of integrating the microbial lactase preparations into functional foods and pharmaceuticals, ultimately paving the way for more effective, accessible, and permanent solutions for individuals with lactose intolerance.

R. Salwan
College of Horticulture and Forestry, Dr YS Parmar University of Horticulture & Forestry, Hamirpur, Himachal Pradesh, India

R. Sharma · R. Kaur · V. Sharma (✉)
University Centre for Research and Development, Chandigarh University, Mohali, Punjab, India

G. Mustafa (✉)
Department of Plant Sciences, Faculty of Biological Sciences, Quaid-i-Azam University, Islamabad, Pakistan
e-mail: mghazala@qau.edu.pk

© The Author(s), under exclusive license to Springer Nature Singapore Pte Ltd. 2026

V. Sharma, R. Salwan (eds.), *Pharmaceutically Relevant Microbial Enzymes*,
https://doi.org/10.1007/978-981-95-4941-2_7

Keywords

Lactose intolerance · Microbial lactase · β-Galactosidase · Enzyme encapsulation · Functional foods

7.1 Introduction

Lactose intolerance affecting approximately 68% of the world's population is one of the most common food-related digestive disorders (Storhaug et al. 2017). The disorder is characterized by the inability to metabolize lactose present in milk and dairy products. The inability to metabolize lactose is due to deficiency in the intestinal enzyme lactase-phlorizin hydrolase, also known as lactase (β-galactosidase). The enzyme hydrolyzes the lactose into its monosaccharide units, i.e., glucose and galactose (Misselwitz et al. 2019). The hydrolyzed lactose then enters into the large intestine where it is fermented by microbiota, leading to the production of gas such as hydrogen (H_2), methane (CH_4), carbon dioxide (CO_2), and short-chain fatty acids. These products often lead to symptoms such as abdominal pain, bloating, flatulence, and diarrhea. The deficiency of lactase varies significantly among different ethnic groups and often associated with the consumption of dairy food. The genetic trait for lactase persistence into adulthood is common in population of Northern Europe having a history of pastoralism and dairy farming. In contrast, regions of Asia, Africa, and South America exhibit a genetically reduced lactase production known as primary adult-type hypolactasia, with prevalence rates often exceeding 90% (Bayless et al. 2017).

The management strategies for lactose intolerance involve the dietary modification and enzyme replacement therapy (ERT). Avoiding lactose-rich diet is one of the most explored methods but also offers drawbacks as dairy products are one of the primary sources of essential nutrients including high-quality protein, calcium (Ca^{2+}), vitamins (D and B_{12}). Therefore, avoiding the dairy-based products can alleviate the risk of nutritional deficiencies, particularly osteoporosis due to inadequate Ca^{2+} intake (Di Rienzo et al. 2013). The alternative measure includes ERT where the oral administration of commercial lactase supplements obtained from microbial sources like *Aspergillus oryzae* or *Kluyveromyces lactis* is done. Although the supplements have been reported for providing temporary relief, the variable and inconsistent effectiveness have been observed as the enzyme activity is compromised by the acidic environment of the stomach as well as by the timing of administration with reference to meal consumed. Therefore, the oral administration is considered as inconvenient and often unreliable solution (Montalto et al. 2006). Therefore, based on the limitations of ERT's oral administration, the alternative approaches, which are effective and reliable, can provide better solutions. With the recent developments in biotechnology, emphasis is now being given to harness the microbial lactases. Microorganisms including bacteria, yeasts, and fungi have been considered as a reservoir of β-galactosidase/lactase. The advantage of exploring microbial sources for lactase includes benefits of easy upscaling, which is also cost-effective and ease of fermentation technologies under a broad range of conditions (Husain 2010). Moreover, the aided advantage of exploring the microbial sources is the advantage of enhancing enzyme stability, catalytic efficiency, thermal stability, and pH tolerance using genetic engineering.

Here, an attempt has been made to cover the significance of microbial lactases as a therapeutic tool for managing lactose intolerance. The chapter also provides an overview of the key advancements from the screening and isolation of novel lactase-producing microorganisms to the optimization of large-scale production. The potential of recombinant DNA technology to engineer the microbial lactases for enhanced enzyme performance and developments in enzyme delivery based on immobilization and microencapsulation are also discussed. The chapter aims to integrate the knowledge of microbial lactase preparations into functional foods and pharmaceuticals, ultimately offering a more robust and accessible solution.

7.2 Lactase Non-persistence and Lactose Intolerance

Lactase non-persistence is a widespread genetic trait that reflects the natural decline of lactase enzyme activity after infancy, while lactose intolerance is the symptomatic manifestation of this physiological condition. Understanding the genetic underpinnings, diverse symptoms, and various diagnostic methods is crucial for accurate identification. Although there is no cure for primary lactose intolerance, effective management strategies, primarily through dietary modifications and the use of enzyme supplements, allow individuals to minimize symptoms and maintain a balanced diet. Lactase non-persistence (LNP), also known as adult-type hypolactasia, is the most common form of lactase deficiency worldwide, affecting 65–70% of adults (Bayless et al. 2017). Lactose, a disaccharide, composed of D-glucose and D-galactose, is unique to mammalian milk and also the primary source of carbohydrate for newborns (Solomons 2002). However, with aging, the lactase production decreases and causes a condition known as lactase non-persistence. The lactase production is genetically regulated and differs significantly among ethnic groups. The term "lactose intolerance" is associated to lactase non-persistence that affects a significant portion of the human population worldwide, which led to digestive discomfort upon consumption of dairy products (Ziarno and Cichońska 2021). The condition arises when the small intestine does not produce enough lactase and hence individual fails to metabolize the lactose (Solomons 2002). The unmetabolized lactose enters into the colon, where it is fermented into gases (such as H_2, CO_2, and CH_4) and short-chain fatty acids (SCFA), which causes symptoms like bloating, abdominal cramps, flatulence, and diarrhea (Bayless et al. 2017; Solomons 2002). The severity of lactose intolerance varies with individuals, the amount of lactose intakes, and the level of lactase deficiency (Heyman 2006). Lactase persistence is a classic example of genetic polymorphism, believed to be associated with the advent of the pastoralist era around 10,000 years ago, when the domestication of milk-producing animals began (Solomons 2002). The domestication of milk-producing animals resulted in lactose to be a part of the human diet after infancy (Solomons

2002). The ability to produce lactase enabled the lactose metabolism, a major sugar in the milk.

The reduction in lactase production after infancy for most mammals including humans is due to the down-regulation of lactase. It is encoded by *LCT* gene, located on chromosome 2q21. In lactase non-persistence individuals, the expression of *LCT* gene gradually decreases with aging and ultimately causes insufficient lactase levels. Contrary to this, the production of lactase is high in lactase persistence (LP) individuals till adulthood. Lactase persistence and non-persistence are considered evolutionary adaptations linked to the development of human civilization, particularly in populations with a long history of pastoralism and dairy consumption (Bayless et al. 2017). The inheritance of lactase persistence has been marked as autosomal dominant trait; therefore, an individual is capable to produce lactase even under heterologous conditions. Further, several single nucleotide polymorphisms (SNPs) in a regulatory region of *LCT* gene (specifically within the MCM6 gene) has been marked (Luo et al. 2024; Yuan et al. 2022). The trait is primarily controlled by regulatory single nucleotide polymorphisms (SNPs) located upstream of *LCT* gene, encoding the lactase enzyme. The substitution at −13,910 C > T variant of *MCM6* gene is responsible for the enhanced expression of *LCT* gene. The substitution for T allele has been strongly linked with lactase persistence in European populations, whereas the ancestral C allele has been linked to lactase non-persistence and lactose intolerance. In addition, other SNPs have been identified from African and Middle Eastern populations, indicating that the lactase persistence evolved independently in multiple regions likely through convergent evolution. Furthermore, the lactase persistence inheritance followed a dominant pattern as only single copy of a persistence-associated allele was sufficient for the trait to be expressed. It has been observed that the distribution of lactase persistence alleles is strongly influenced by the history of dairy farming and milk consumption, illustrating a clear example of "gene-culture coevolution" in human evolution (Swallow 2003; Liebert et al. 2017). Comparative analyses of lactase non-persistent individuals have shown that higher milk intake leads to significant alterations in the gut microbiota. For example, increased milk consumption is associated with a higher abundance of *Bifidobacterium* and a reduction in Prevotella, along with elevated levels of indolepropionate and decreased levels of branched-chain amino acid metabolites (Luo et al. 2024). Moreover, the influence of host genetic factors on the human microbiome has been explored. A genome-wide association study identified an age-dependent interaction affecting the abundance of *Bifidobacterium* (Kurilshikov et al. 2021).

7.3 Biochemical Characteristics and Classification of Lactase

Lactase, also referred to as lactase-phlorizin hydrolase (LPH), is classified under Enzyme Commission number EC 3.2.1.108. LPH, along with glycosylceramidase (EC 3.2.1.62), is exclusively expressed in the small intestine and located on the brush border membrane of enterocytes. Both enzymatic activities are functions of the same protein, lactase-phlorizin hydrolase (LPH), which hydrolyzes lactose into glucose and galactose (Naim et al. 1994; Skovbjerg et al. 1981). Structurally, consisting of 1927 residues, lactase consists of a four-fold internal homology which represents four distinct domains (I-IV). The enzyme is translated as a large precursor polypeptide (pre-pro-LPH) that post-translationally modified (Naim et al. 1994; Skovbjerg et al. 1981; Troelsen 2005). Lactase is an ecto-enzyme and its catalytic domains are located extracellularly, anchored to the microvillar membrane through its C-terminally located hydrophobic stretch (Naim et al. 1994; Troelsen 2005). During transport, the precursor protein is transported from the endoplasmic reticulum to the Golgi apparatus and then ultimately to the brush border. During translocation, the precursor polypeptide also undergoes N, O-glycosylation and proteolytic cleavage. The pro-peptide from domains I and II is cleaved off. The pro-peptide is considered to act as an intramolecular chaperone that assists protein folding prior to its final destination (Naim et al. 1994; Troelsen 2005). The mature lactase is composed of two identical 160 kDa polypeptide units (Troelsen 2005). The domains III and IV share homology to glycosyl hydrolase family 1 (GH1) and domain IV contains the active site that is involved for catalysis. The domain III is responsible for phlorizin hydrolase activity (Arribas et al. 2000). Compared to mammals, the microbial β-galactosidases typically exist as tetramers or dimers with size of 116 to 540 kDa (Juers et al. 2001).

Lactase, expressed in the brush border membrane of the small intestine in mammals, has also been reported for microbial sources such as *Aspergillus oryzae*, *Bacillus circulans*, *Escherichia coli*, and *Kluyveromyces lactis* (Panesar et al. 2007). The optima temperature and pH for different microbial lactases lies in the range 30–70 °C and 4.0 to 7.5, respectively. Similar to other glycosyl hydrolases, lactase from fungal sources prefers acidic conditions, whereas the bacterial sources act under neutral to alkaline conditions. Due to pH and thermal stability, the immobilized microbial lactases are preferred for industrial use for repeated batch operations (Gekas and López-Leiva 1985; Parmar et al. 2021). Compared to human lactases which are known for their high specificity for lactose, the enzymes from microbial sources can act on a broad range of β-galactoside substrates (Panesar et al. 2007). The metal ions such as Hg^{2+}, Cu^{2+}, and Ag^{+} suppress the enzyme activity, whereas Mg^{2+} and Mn^{2+} are considered for their positive effect on enzyme activity (Mahoney 1998).

7.4 Microbial Sources of Lactases: Recent Developments in Characterization, Purification, and Cloning

Lactases (β-galactosidases) from microbial sources are of critical importance due to their broad substrate specificity. The microbial lactases are used in dairy industry for producing lactose-free products and also as ERT for individuals with lactose intolerance. The microbial strains have been identified and explored for novel lactases with improved activity. The genetic engineering strategies such as gene cloning and site directed mutagenesis have been used for enhancing the stability and other biochemical characteristic of lactases. The exploration of microbial sources has already led to the identification of enzymes that are active under a broad pH and temperature ranges and with better catalytic efficiency. For instance, *Gal42*, a β-galactosidase, has been cloned and characterized from *Bacillus* sp. BY02, isolated from the marine habitat. The enzyme effectively hydrolyzes the lactose, indicating its potential in food applications (Wang et al. 2021). In 2022, lactase from *B. licheniformis* ALSZ2 with a broad temperature range (30–60 °C) and pH optima of 7 found suitable as pharmaceutical supplement for lactose intolerance (Movahedpour et al. 2022). On the other side, the lactases active at low temperature have been explored from psychrophilic microorganisms for dairy industry. Contrary to this, lactases stable at elevated temperatures from thermophilic bacteria are being used to improve the rate of reaction and managing the contamination risks (Schallmey et al. 2004). The lactases from extremophilic microorganisms have been investigated for their stability and activity under harsh industrial conditions (Sharma et al. 2020).

The microbial lactase encoding genes have been cloned into heterologous hosts such as *Escherichia coli*. The lactase genes from *Bacillus licheniformis* (Schallmey et al. 2004; Wang et al. 2021) and *Bifidobacterium* species (Zhang et al. 2022) have been successfully cloned for large-scale industrial applications. Additionally, the chimeric lactase has developed by fusing with a cellulose-binding module (CBM) of *Anaerocellum thermophilum*. The immobilized recombinant lactase showed continuous hydrolysis of lactose even at high temperature. Furthermore, the advancements in sequencing and metagenomics technologies have not only accelerated the discovery but also expanded the repertoire of novel lactase genes even from uncultured microbial communities.

7.5 Enhancing the Stability of Microbial Lactases

The stability of microbial lactases is critical for their practical application, especially under extremes of industrial conditions involving varying temperature, pH, and the presence of inhibitors. In addition, immobilization of lactases has been explored for improving the stability, reusability, and even catalytic efficiency of microbial lactases. Various polymeric and nanomaterials have been used for immobilizing microbial lactases. For example, lactase from *A. oryzae* has been effectively immobilized on cellulose acetate nanofibers, resulting in improved thermal stability and reusability (Deng et al. 2024). The immobilized lactase onto modified mesoporous silica/titania and chitosan retained 90% of its initial enzyme activity after repeated usage. The gold nanoparticles-based preparations enhanced the lactases stability and performance under a wide range of temperature and pH (Deng et al. 2024). In addition, encapsulation of calcium alginate has been used. Ingredients such as trehalose or guar gum are also used for encapsulation to enhance and preserve the thermal and mechanical stresses (Traffano-Schiffo et al. 2017). Besides this, the directed evolution and rational design approaches have been used to engineer lactases with enhanced stability, pH optima, and improved resistance to degradation (Pérez-Vega et al. 2023).

7.6 Lactase-Deficiency and Therapeutics

The physiological inability to produce sufficient lactase at the small intestine's brush border is known as lactase deficiency (LD). The prevalence of lactase deficiency due to non-persistent lactase expression in the small intestinal mucosa varies widely across populations, ranging from approximately 2% in Caucasians to over 90% in many Asian and Native American groups (Sun et al. 2015). The secondary lactase deficiency results from damage to small intestinal mucosa due to infections in individuals who otherwise have persistent lactase expression (Sun et al. 2015). In newborns, rare congenital lactase deficiency (CLD) causes disaccharide intolerance, absorption issues, and watery diarrhea upon lactose feeding. The mutation in lactase phlorizin hydrolase (LPH) is responsible for CLD. In 2015, a comparative study of infants with CLD revealed that majority of them have hypercalcemia and medullary nephrocalcinosis and the symptoms improved on a lactose-free diet. Even though the mechanism of hypercalcemia was unclear, it was observed that it led to metabolic acidosis or lactose-related effects (Sun et al. 2015). The genetic base of CLD patients revealed that the homozygous mutation in the lactase encoding gene resulted in a truncated and inactive enzyme that fails to reach the cell surface during protein translocation (Behrendt et al. 2009). The symptoms in CLD children are visible from beginning of their birth. The genetic cause of adult-type hypolactasia has been mapped to variations in C/T at 13910 and G/A at 22018 that led to reduction in lactase activity (Behrendt et al. 2009). Post-weaning data revealed that only about one-third of people develop symptoms of lactose intolerance due to reduced lactase activity (Rasinkangas et al. 2022).

7.7 Lactase-Based Treatment Strategies for Lactose Intolerance

Lactose intolerance, a common digestive disorder, is characterized by the inability to fully digest lactose. Its diagnosis relies on several methods including the lactose tolerance test, which measures blood glucose levels after lactose consumption, and the hydrogen breath test, which detects hydrogen gas produced by bacterial fermentation of undigested lactose in the colon (Leis et al. 2020; Mattar et al. 2012). Genetic testing can also identify specific variants associated with lactase persistence or non-persistence (Solomons 2002). Although the disease is not life-threatening, the symptoms of lactose intolerance such as bloating, cramping, and flatulence may significantly affect individual's life largely due to production of CO_2, H_2, and CH_4 of lactose fermentation (Solomons 2002).

Lactase supplements are used to treat lactose intolerance. The supplements such as LACTAID Fast Act Caplets are used as external source of lactase to metabolize the lactose in the gastrointestinal tract which also prevents undigested lactose to reach the colon and cause symptoms. The supplements are also available as tablets, liquids, and chewable for easy consumption. While safe, the lactase supplements also come with limitations. Although considered as safe but the efficacy varies from individual to individual and in some cases, symptoms can emerge if utilized lactose remains. Further, determining the optimal dose is challenging as it varies from individual to individual due to lactose consumption and individual physiological responses. In a few cases, mild side effects such as stomach discomfort, nausea, or bloating and even allergic reactions to inactive ingredients have been observed. The variations in the efficacy and optimal dose determination for lactase supplements underscore a significant challenge addressing the lactose intolerance. The complexity of lactose tolerance and stability of enzyme in stomach demand the need for alternative formulations that are better and can withstand harsh gastrointestinal environment. The role of microbial lactases varies with biochemical stability of lactases and has been utilized to alleviate lactose intolerance in dairy products for low-lactose or lactose-free alternative (Grosová et al. 2008; Harju et al. 2012; Saqib et al. 2017). Such lactose-free alternative dairy products avoid triggering lactose intolerance symptoms but increase sweetness that can alter the product's taste profile (Dekker et al. 2019). Lactose-free diets in individuals with lactose intolerance may lead to deficiencies in nutrients such as calcium (Ca^{2+}), phosphorus (P), and vitamin D, which can negatively affect bone health (Mesa 2020). Besides benefits of lactases, preserving enzyme activity during processing, storage, and gastrointestinal path is challenging due to extreme conditions. The extreme conditions such as temperature fluctuations, high acidity, and high protease activity in stomach negatively affect lactase activity. The encapsulation of lactase using food-grade materials has emerged as attractive strategy to enhance the stability of lactase (Perry and McClements 2020). For encapsulation, different methods such as nanomaterials, emulsions, and gels are being explored. The success of these delivery methods is based on the retention, stability, and releasing time of protein after consumption.

7.8 Developments in Lactase Delivery Systems

The highly acidic environment of stomach is a major challenge for oral lactase supplements as it can deactivate the enzyme. Form this, novel drug delivery systems have been developed to protect the enzyme and ensure effective and functional delivery. A combination of encapsulation and nanotechnology is at the forefront of these advancements aiming for improved stability and targeted release of lactase. Encapsulation of lactase within nanomaterials such as zein/pectin capsules has been found promising. These enzyme-matrices can be mixed with dairy products particularly milk for retaining lactase activity during storage for selective release of enzyme in the intestines after ingestion. The efficiency of 93% encapsulation of zein/pectin capsules has been achieved. Further, in vitro digestion studies indicated that encapsulated enzyme hydrolyzes 100% of lactose in whole milk and 89.4% in skim milk, which is significantly higher than equivalent units of free lactase. The core-shell of nanoparticles offers distinct advantages and stability with controlled-release. In this context, the core-shell electrospinning technique has been used for the encapsulation of lactase within an aqueous core protected. The non-biodegradable shell enhanced the enzyme stability that reduces early diffusion limitations and improves reusability. The enzyme in core shell was stable for almost 28 days and retained enzyme activity even after 3 months of storage at 4 °C.

The traditional method involves eliminating dietary lactose by immobilizing lactase to break down lactose in dairy products, which can be used industrially to produce low-lactose or lactose-free dairy products (Grosová et al. 2008; Harju et al. 2012; Saqib et al. 2017). The lactase formulations are packaged as acid-proof tablets to avoid the damage acid in stomach. The developments in lactase delivery are based on our understanding of gastrointestinal physiology and advancements in drug delivery approaches that use simple acid-resistant coatings to precisely control enzyme release for maximum activity and low off-target effects. Using our collective understanding, the enzyme supplementation has emerged from a crude "dump-and-digest" approach to a comparatively refined, targeted therapeutic intervention. Furthermore, efforts are also being made to explore the synergistic approaches for optimum performance. The immobilized lactase in combination with glucose oxidase and catalase has been investigated to decrease the feedback inhibition by glucose, after lactose hydrolysis, and combat the toxicity of hydrogen peroxide (H_2O_2). Such developments on using multi-enzyme systems help in maintaining optimal lactase performance by involving glucose oxidase to metabolize glucose and catalase to mitigate the potential oxidative stress. Further, such approaches also assist in optimizing the biochemical microenvironment that is useful in maintaining the environment and functionality of the enzyme with better efficacy for diverse applications. In recent years, direct ingestion of lactase as a dietary supplement has become a research focus to ensure balanced nutrition for lactose-intolerant individuals (Ferreira-Lazarte et al. 2018; McClements 2004; Mesa 2020). Lactase or β-galactosidase (β-Gal), approved by the USFDA, is widely used in the dairy industry to produce lactose-free milk, which aids lactose-intolerant individuals and promotes gut health due to increased galactooligosaccharides (Saqib et al. 2017; Harju et al. 2012). Studies also reveal that β-Gal after thermal milk processing is more cost-effective and reduces nutrient loss from Maillard reactions (Tossavainen and Kallioinen 2008). However, using free β-Gal comes with limitations including low stability, poor reusability, and high cost. Therefore, immobilized β-Gal has been used to enhance the thermal and pH stability, storage and reusability (Gonçalves et al. 2019). The immobilization also enables the continuous lactose hydrolysis. However, it also brings challenges of microbial growth at neutral pH and protein adsorption on enzyme surfaces (Damin et al. 2021). Using adsorption-based physical interactions such as van der Waals and hydrogen bonding between β-Gal and carriers like cellulose acetate fibers or hydroxypropyl methylcellulose, a reduction of up to 89% lactose with high reusability has been achieved (Harju 2004; Silva et al. 2020; Shafi et al. 2021).

The encapsulation of β-Gal in polymer hydrogels like PVA or natural polymers (e.g., alginate, chitosan) has resulted in improved enzyme activity and shelf-life.

The PVA-immobilized β-Gal has been successfully used and can be stored for 14 months at 4 °C (Grosová et al. 2008). Further, the covalently immobilized enzyme to carriers has been reported for enhancing the stability and reusability of the enzyme (Damin et al. 2021). Matrices or carriers such as chitosan, graphite, oxides, and other nanomaterials have been found promising. For instance, β-Gal immobilized onto modified chitosan containing glutaraldehyde or epoxy groups has been reported for high activity and enhanced stability (Makowski et al. 2007; de Freitas et al. 2020). The improvement in enzyme loading, enzyme attachment, and stability has been used to improve the performance. The adsorption and cross-linking of β-Gal on concanavalin A-Celite 545 led to increased lactose hydrolysis (Ansari and Husain 2012), whereas the crosslinking with metal ions resulted in improved thermal stability (Ospina et al. 2019). However, concerns such as alteration of enzyme structure and specificity, decrease in activity, and formation of unwanted side-products have been raised. Therefore, the application of covalent and cross-linking methods enhances lactase performance, but it does need assessment of any side effects.

7.9 Modulating the Gut Microbiome for Lactose Digestion

Lactase deficiency is the outcome of human physiological inability to produce sufficient lactase at the small intestine's brush border (Sun et al. 2015). Lactase deficiency suppresses the ability to digest or absorb lactose in the small intestine. Among various therapeutic strategies used to address the gut microbiome, probiotics, prebiotics, synbiotics, and postbiotics have emerged as effective approaches for enhancing lactose digestion and alleviating symptoms of lactose intolerance. The presence of prebiotics such as GOS specifically favors the growth of beneficial lactose-fermenting gut microbes such as *Bifidobacterium* and *Lactobacillus*. These microorganisms are potential source of β-gal and hence partially compensate the human lactase deficiency. Therefore, the stimulation of beneficial strains can improve lactose tolerance and alleviate the associated symptoms. Studies revealed that even a regular intake of lactose in lower dose acts as a prebiotic that promotes the lactose-metabolizing microbes and ultimately improves lactose tolerance (Janssen-Duijghuijsen et al. 2024). Consistent efforts have been made to improve microbial gut flora that is responsible for boosting health via modulating immune responses and other pathways.

Among probiotic strains, *Lactobacillus* and *Bifidobacterium* have been reported for producing lactase that can directly metabolize lactose in the intestinal lumen. The lactose metabolism by these probiotic strains catalyzes the production of short-chain fatty acids that ultimately enhances the gut health and even microbial balance. Strains such as *L. acidophilus*, *B. longum*, *L. rhamnosus* GG, *L. bulgaricus*, and *Streptococcus thermophilus* have been reported for improving lactose digestion and alleviating symptoms like bloating, abdominal cramps, and diarrhea. The experimental data revealed that β-galactosidase-producing probiotic strains such as *S. thermophilus* and *B. longum* facilitate lactose maldigestion. Even synbiotic approaches use a combination of prebiotics and probiotics which have been explored against cow's milk allergy in infants. It has been observed that modulation of gut microbiota can be used to reduce infections and promote tolerance. Still, the underlying mechanism of microbiome-mediated modulation needs to be investigated.

Recent research is transforming our understanding of lactose intolerance management, shifting focus from a reliance on external enzyme supplements to the remarkable adaptability of the gut microbiome. A pivotal study by Janssen-Duijghuijsen et al. (2024) demonstrated healthy Asian adults with the LNP-genotype experienced a significant rise in beneficial *Bifidobacterium* and a twofold increase in fecal β-galactosidase activity leading to improved lactose metabolism. This suggests that while human lactase deficiency is permanent, the microbial ecosystem in the colon can be trained to compensate, offering a more sustainable and internally regulated solution with potential benefits for gut health and paving the way for personalized probiotic interventions. This contrasts sharply with long-standing evidence that endogenous intestinal lactase activity is not influenced by dietary lactose intake. Similar studies consistently showed no increase in lactase activity after prolonged lactose consumption with age, not diet, being the primary factor affecting lactase expression (Huang and Bayless 1968). Therefore, effective management of lactose intolerance increasingly points towards harnessing the adaptive capacity of the gut microbiome rather than solely relying on enzyme replacement to support both digestive health and overall well-being.

7.10 Conclusions and Future Perspectives

Lactase deficiency poses a significant global health challenge particularly given the high prevalence in Asian and Native American populations. While the physiological inability to produce sufficient lactase is a core issue, advancements in managing lactose intolerance are moving beyond ERT. Emerging strategies focus on modulating the gut microbiome through targeted interventions like prebiotics, probiotics, and synbiotics which encourage the growth and activity of lactose-fermenting bacteria. The modulation of gut microbiota such as *Bifidobacterium* with β-galactosidase activity and lactose intake offers a more sustainable, therapeutics measures compared to traditional exogenous lactase supplementation. However, the endogenous intestinal lactase activity is largely insufficient to dietary lactose levels and often demands alternative strategies. Recently, the significant efforts have been made in lactase delivery systems such as using encapsulation and nanotechnology, protecting the enzyme from gastric degradation, and ensuring its targeted release in the small intestine. These developments in delivery methods and explorations of multi-enzyme systems are being explored for precise and effective therapeutic interventions for lactose intolerance. The therapeutic measures for lactose intolerance involve a dynamic interplay between microbiome engineering and advanced enzyme delivery technologies. The consistent efforts into specific microbial strains and tailored prebiotic interventions may be useful. Therefore, developing synbiotic formulations that can facilitate the growth of lactose-metabolizing bacteria is needed. In parallel, the developments in encapsulation methods are needed for controlled release and enhanced stability under gastrointestinal conditions. Furthermore, integrating diagnostic tools that rapidly assess an individual's lactase activity and gut microbiome profile will be crucial for guiding these personalized therapeutic approaches. The holistic perspective will promise to transform the management of LD from a restrictive dietary challenge into a manageable aspect of digestive wellness.

References

Ansari SA, Husain Q (2012) Covalent immobilization of β-galactosidase on concanavalin A-Celite 545 for lactose hydrolysis. Int J Biol Macromol 50(4):1014–1020

Arribas JC, Herrero AG, Martín-Lomas M, Cañada FJ, He S, Withers SG (2000) Differential mechanism-based labeling and unequivocal activity assignment of the two active sites of intestinal lactase/phlorizin hydrolase. Eur J Biochem 267(24):6996–7005

Bayless TM, Brown E, Paige DM (2017) Lactase non-persistence and lactose intolerance. Curr Gastroenterol Rep 19(5):23

Behrendt M, Keiser M, Hoch M, Naim HY (2009) Impaired trafficking and subcellular localization of a mutant lactase associated with congenital lactase deficiency. Gastroenterology 136(7):2295–2303

Damin FM, Rovani B, Treichel H (2021) Immobilization of β-galactosidase for lactose hydrolysis: a review. Enzym Microb Technol 143:109720

de Freitas ML, de Matos CP, da Silva LG, de Assis RL (2020) Covalent immobilization of β-galactosidase on glutaraldehyde-activated chitosan and its application in lactose hydrolysis. Biocatal Biotransformation 38(4):282–290

Dekker PJ, Koenders D, Bruins MJ (2019) Lactose-free dairy products: market developments, production, nutrition and health benefits. Nutrients 11(3):551

Deng Z, Deng Q, Li B, Li J, Jung S, Cho NJ, Liang H (2024) Strategies for lactase immobilization and delivery to relieve lactose intolerance. Trends Food Sci Technol 143:104244

Di Rienzo T, D'angelo G, D'aversa F, Campanale MC, Cesario V, Montalto M, Gasbarrini A, Ojetti V (2013) Lactose intolerance: from diagnosis to correct management. Eur Rev Med Pharm Sci 16:17

Ferreira-Lazarte A, Moreno FJ, Villamiel M (2018) Application of a commercial digestive supplement formulated with enzymes and probiotics in lactase non-persistence management. Food Function 9(9):4642–4650

Gekas V, López-Leiva M (1985) Hydrolysis of lactose: a literature review. Process Biochem 20(1):2–12. ref. 157

Gonçalves MM, da Silva TL, Guimarães JH, de Souza RR, Converti A (2019) Immobilization of lactase: a review. Process Biochem 84:145–156

Grosová Z, Rosenberg M, Rebroš M, Šipocz M, Sedláčková B (2008) Entrapment of β-galactosidase in polyvinylalcohol hydrogel. Biotechnol Lett 30(4):763–767

Harju M (2004) Lactose hydrolysis. In: Enzymatic conversion of biomass for fuels and chemicals. Springer, pp 209–224

Harju M, Kallioinen M, Tossavainen O (2012) Lactose hydrolysis and other uses of β-galactosidases in the dairy industry. In: Enzymes in food technology, 2nd edn. Wiley, pp 605–621

Heyman MB (2006) Committee on nutrition. Lactose intolerance in infants, children, and adolescents. Pediatrics 118(3):1279–1286

Huang SS, Bayless TM (1968) Milk and lactose intolerance in healthy Orientals. Science 160(3823):83–84

Husain Q (2010) β galactosidases and their potential applications: a review. Crit Rev Biotechnol 30(1):41–62
Janssen-Duijghuijsen L, Looijesteijn E, van den Belt M, Gerhard B, Ziegler M, Ariens R, Geurts J (2024) Changes in gut microbiota and lactose intolerance symptoms before and after daily lactose supplementation in individuals with the lactase nonpersistent genotype. The Am J Clin Nutr 119(3):702–710
Juers DH, Heightman TD, Vasella A, McCarter JD, Mackenzie L, Withers SG, Matthews BW (2001) A structural view of the action of *Escherichia coli* (lac Z) β-galactosidase. Biochemistry 40(49):14781–14794
Kurilshikov A, Medina-Gomez C, Bacigalupe R, Radjabzadeh D, Wang J, Demirkan A, Le Roy CI, Raygoza Garay JA, Finnicum CT, Liu X, Zhernakova DV, Bonder MJ, Hansen TH, Frost F, Rühlemann MC, Turpin W, Moon JY, Kim HN, Lüll K, Barkan E, Shah SA, Fornage M, Szopinska-Tokov J, Wallen ZD, Borisevich D, Agreus L, Andreasson A, Bang C, Bedrani L, Bell JT, Bisgaard H, Boehnke M, Boomsma DI, Burk RD, Claringbould A, Croitoru K, Davies GE, van Duijn CM, Duijts L, Falony G, Fu J, van der Graaf A, Hansen T, Homuth G, Hughes DA, Ijzerman RG, Jackson MA, Jaddoe VWV, Joossens M, Jørgensen T, Keszthelyi D, Knight R, Laakso M, Laudes M, Launer LJ, Lieb W, Lusis AJ, Masclee AAM, Moll HA, Mujagic Z, Qibin Q, Rothschild D, Shin H, Sørensen SJ, Steves CJ, Thorsen J, Timpson NJ, Tito RY, Vieira-Silva S, Völker U, Völzke H, Võsa U, Wade KH, Walter S, Watanabe K, Weiss S, Weiss FU, Weissbrod O, Westra HJ, Willemsen G, Payami H, Jonkers DMAE, Arias Vasquez A, de Geus EJC, Meyer KA, Stokholm J, Segal E, Org E, Wijmenga C, Kim HL, Kaplan RC, Spector TD, Uitterlinden AG, Rivadeneira F, Franke A, Lerch MM, Franke L, Sanna S, D'Amato M, Pedersen O, Paterson AD, Kraaij R, Raes J, Zhernakova A (2021) Large-scale association analyses identify host factors influencing human gut microbiome composition. Nat Genet 53(2):156–165
Leis R, de Castro MJ, de Lamas C, Picáns R, Couce ML (2020) Effects of prebiotic and probiotic supplementation on lactase deficiency and lactose intolerance: a systematic review of controlled trials. Nutrients 12(5):1487
Liebert A, López S, Jones BL, Montalva N, Gerbault P, Lau W, Thomas MG, Bradman N, Maniatis N, Swallow DM (2017) World-wide distributions of lactase persistence alleles and the complex effects of recombination and selection. Hum Genet 136(11):1445–1453
Luo K, Chen GC, Zhang Y, Moon JY, Xing J, Peters BA, Usyk M, Wang Z, Hu G, Li J, Selvin E, Rebholz CM, Wang T, Isasi CR, Yu B, Knight R, Boerwinkle E, Burk RD, Kaplan RC, Qi Q (2024) Variant of the lactase LCT gene explains association between milk intake and incident type 2 diabetes. Nat Metab 6(1):169–186
Mahoney RR (1998) Galactosyl-oligosaccharide formation during lactose hydrolysis: a review. Food Chem 63(2):147–154
Makowski P, Piatek D, Szczęsna A (2007) Immobilization of β-galactosidase on chitosan modified with epoxy groups. J Mol Catal B Enzym 49(1–4):15–20
Mattar R, de Campos Mazo DF, Carrilho FJ (2012) Lactose intolerance: diagnosis, genetic, and clinical factors. Clin Exp Gastroenterol 5:113–121
McClements DJ (2004) Food emulsions: principles, practices, and techniques. CRC press
Mesa MD (2020) Lactase supplements for lactose intolerance: an update. Nutrients 12(7):2056
Misselwitz B, Butter M, Verbeke K, Fox MR (2019) Update on lactose malabsorption and intolerance: pathogenesis, diagnosis and clinical management. Gut 68(11):2080–2091
Montalto M, Curigliano V, Santoro L, Vastola M, Cammarota G, Manna R, Gasbarrini A, Gasbarrini G (2006) Management and treatment of lactose malabsorption. World J Gastroenterol 12(2):187
Movahedpour A, Ahmadi N, Ghalamfarsa F, Ghesmati Z, Khalifeh M, Maleksabet A, Shabaninejad Z, Taheri-Anganeh M, Savardashtaki A (2022) β-Galactosidase: from its source and applications to its recombinant form. Biotechnol Appl Biochem 69(2):612–628
Naim HY, Jawb R, Naim H, Sambrook JF, Gething MJ (1994) The pro region of human intestinal lactase-phlorizin hydrolase. J Biol Chem 269:26933–26943
Ospina S, Pérez F, Rodríguez M (2019) Cross-linking of β-galactosidase with metal ions for enhanced thermal stability. Enzym Microb Technol 122:1–7

Panesar R, Panesar PS, Singh RS, Kennedy JF, Bera MB (2007) Production of lactose-hydrolyzed milk using ethanol permeabilized yeast cells. Food Chem 101(2):786–790

Parmar P, Lopez-Villalobos N, Tobin JT, Murphy E, Buckley F, Crowley SV, Kelly AL, Shalloo L (2021) Effect of temperature on raw whole milk density and its potential impact on milk payment in the dairy industry. Int J Food Sci Technol 56(5):2415–2422

Pérez-Vega MJ, López-Gallego F, Alcalde M (2023) Engineering β-galactosidases: from structure to applications through directed evolution and rational design. Biotechnol Adv 63:108152

Perry SL, McClements DJ (2020) Recent advances in encapsulation, protection, and oral delivery of bioactive proteins and peptides using colloidal systems. Molecules 25(5):1161

Rasinkangas P, Forssten SD, Marttinen M, Ibarra A, Bothe G, Junnila J, Uebelhack R, Donazzolo Y, Ouwehand AC (2022) *Bifidobacterium animalis* subsp. *lactis* Bi-07 supports lactose digestion *in vitro* and in randomized, placebo-and lactase-controlled clinical trials. Am J Clin Nutr 116(6):1580–1594

Saqib S, Khan MR, Muhammad H (2017) β-galactosidase: a comprehensive review of its applications. Biotechnol Prog 33(4):842–859

Schallmey M, Singh A, Ward OP (2004) Developments in the use of *Bacillus* species for industrial production. Can J Microbiol 50(1):1–7

Shafi AM, El-Sayed EA, Salem SH (2021) Immobilization of lactase on hydroxypropyl methylcellulose for lactose hydrolysis. J Appl Polym Sci 138(1):49767

Sharma AK, Lourembam B, Bisht B (2020) New approach to plant-based Milk alternatives-a review. Ind J Pure App Biosci 8(3):130–133

Silva ML, Souza JC, Dutra PV (2020) Adsorption of β-galactosidase on cellulose acetate fibers for lactose hydrolysis. Appl Biochem Biotechnol 191(2):642–655

Skovbjerg H, Sjostrom H, Noren O (1981) Purification and characterisation of amphiphilic lactaselphiorizin hydrolase from human small intestine. Eur J Biochem 114:653–661

Solomons NW (2002) Fermentation, fermented foods and lactose intolerance. Eur J Clin Nutr 56(4):S50–S55

Storhaug CL, Fosse SK, Fadnes LT (2017) Country, regional, and global estimates for lactose malabsorption in adults: a systematic review and meta-analysis. Lancet Gastroenterol Hepatol 2(10):738–746

Sun J, Lin J, Parashette K, Zhang J, Fan R (2015) Association of lymphocytic colitis and lactase deficiency in pediatric population. Pathol Res Pract 211(2):138–144

Swallow DM (2003) Genetics of lactase persistence and lactose intolerance. Annu Rev Genet 37(1):197–219

Tossavainen O, Kallioinen M (2008) Lactose hydrolysis in milk and whey products: technological and nutritional aspects. Int Dairy J 18(9):896–905

Traffano-Schiffo MV, Castro-Giraldez M, Fito PJ, Santagapita PR (2017) Encapsulation of lactase in Ca (II)-alginate beads: effect of stabilizers and drying methods. Food Res Int 100:296–303

Troelsen JT (2005) Adult-type hypolactasia and regulation of lactase expression. Biochim Biophys Acta Gen Subjects 1723(1–3):19–32

Wang L, Wang YW, Tan JT, Yan J, Wu Y, Wang XM, Yang WZ, Qian JH (2021) Efficacy and safety of lactase additive in preterm infants with lactose intolerance: a prospective randomized controlled trial. J Contemp Pediatr 23(7):671–676

Yuan S, Sun J, Lu Y, Xu F, Li D, Jiang F, Wan Z, Li X, Qin LQ, Larsson SC (2022) Health effects of milk consumption: phenome-wide Mendelian randomization study. BMC Med 20(1):455

Zhang W, Wang S, Xing Y, Wang H, Fu B, Long M, Cao J (2022) Clinical efficacy of probiotics on feeding intolerance in preterm infants: a systematic review and meta-analysis. Transl Pediatr 11(2):229

Ziarno M, Cichońska P (2021) Lactic acid bacteria-fermentable cereal-and pseudocereal-based beverages. Microorganisms 9(12):2532

Structure and Function of Microbial Glutaminases and Their Therapeutics Applications

8

Vivek Sharma, Rhydum Sharma, and Ewa B. Moliszewska

Abstract

Glutaminase, reported from various microorganisms such as bacteria, yeasts, and fungi, primarily catalyzes the hydrolysis of the γ-amido bond of L-glutamine to yield L-glutamic acid. Certain glutaminases also facilitate a γ-glutamyl transfer reaction, which is valuable for synthesizing bioactive compounds such as theanine, a flavor-enhancing amino acid naturally present in green tea. Owing to this unique functionality, microbial glutaminases are widely employed as flavor enhancers in the food industry. Despite these promising applications, a significant knowledge gap remains in understanding the structural characteristics of microbial glutaminases, particularly their primary and secondary structures. Additionally, this chapter explores the therapeutic applications of microbial glutaminases, examining their role in monotherapy and combination strategies for cancer treatment, as well as their potential in other diseases. This chapter attempts to provide insights into their physicochemical properties, thereby contributing to a deeper understanding of the structure–function relationship of microbial glutaminases and paving the way for their therapeutic and industrial applicability.

Keywords

Microbial glutaminase · Therapeutics · Enzyme structure–function · Health and nutrition · Biotechnology innovation · Sustainable food systems

---

V. Sharma · R. Sharma (✉)
University Centre for Research and Development Chandigarh University, Mohali, Punjab, India

E. B. Moliszewska (✉)
Institute of Environmental Engineering and Biotechnology, University of Opole, Opole, Poland
e-mail: ewamoli@uni.opole.pl

© The Author(s), under exclusive license to Springer Nature Singapore Pte Ltd. 2026

V. Sharma, R. Salwan (eds.), *Pharmaceutically Relevant Microbial Enzymes*,
https://doi.org/10.1007/978-981-95-4941-2_8

8.1 Introduction

Enzymes derived from microbial sources have garnered significant attention for their diverse therapeutic applications, particularly in the treatment of various diseases including cancer, infections, and metabolic disorders (Vanderstocken et al. 2022). This burgeoning interest stems from their cost-effective production, ease of large-scale manufacturing, and reduced ethical concerns compared to enzymes sourced from animals (Vanderstocken et al. 2022). Furthermore, the catalytic efficiency and specificity inherent to enzymes make them exceptionally potent therapeutic agents, capable of performing biochemical conversions that small molecule drugs often cannot (Vellard 2003). A prime example of enzyme therapeutics in oncology is L-asparaginase, which has gained FDA approval for treating acute lymphoblastic leukemia by depleting asparagine, an essential nutrient for cancerous cells (Vellard 2003; Pokrovsky et al. 2022). Similarly, glutaminases, which catalyze the hydrolysis of L-glutamine, represent another class of enzymes with considerable therapeutic potential due to the frequent L-glutamine dependency observed in many rapidly proliferating tumor cells (Endicott et al. 2021). This dependency makes glutamine a pivotal target for antineoplastic strategies, as its depletion can lead to selective cytotoxicity in cancer cells while sparing normal, non-transformed cells (Giurini et al. 2024). This therapeutic strategy, known as glutamine deprivation, aims to exploit the metabolic vulnerabilities of cancer cells by inhibiting their primary anaplerotic pathway and nucleotide synthesis (Martínez-Rodríguez et al. 2020). Nitrogen acquisition and utilization are tightly regulated processes essential for cell growth and proliferation, as nitrogen limitations can impair the synthesis of crucial biomolecules such as proteins and nucleic acids (Kurmi and Haigis 2020). The strategic targeting of glutamine metabolism by glutaminases therefore presents a promising avenue for therapeutic intervention, particularly in cancers exhibiting elevated glutamine addiction (Moldovan et al. 2023). Given this therapeutic potential, understanding the structural and functional characteristics of microbial glutaminases is crucial for optimizing their efficacy and expanding their clinical utility (Vanderstocken et al. 2022).

Microbial glutaminase is an enzyme of growing industrial significance, particularly valued for its role in catalyzing the hydrolysis of L-glutamine. This reaction deamidates L-glutamine into L-glutamic acid, a key compound recognized for its highly savory taste. This unique functional property makes glutaminase a potent flavor-enhancing agent in the food industry. Beyond this primary function, certain glutaminases can also catalyze a γ-glutamyl transfer reaction, enabling the synthesis of other beneficial compounds, such as theanine, which enhances the taste of green tea. Microbial sources of glutaminase from bacteria, fungi, and yeasts are highly preferred over animal sources due to their superior stability, greater affordability, and simplified production process (Al-Thawadi 2011). The versatility of microbial glutaminases extends beyond their well-known role in food flavor. These enzymes possess considerable potential in the pharmaceutical industry, where they are being explored for their anticancer and antiviral properties. Additionally, they are valuable components in biosensors for detecting L-glutamine and are instrumental in producing nutraceuticals like theanine. This broad range of applications highlights the importance of continued research into microbial glutaminases, not only to fully explore their current potential but also to identify new avenues for their use. Microbial glutaminases are classified as L-glutamine amidohydrolases with the Enzyme Commission number EC 3.5.1.2. This classification places them in the hydrolase group, specifically enzymes that cleave non-peptide carbon–nitrogen bonds, facilitating the conversion of L-glutamine to L-glutamic acid and ammonia. This catalytic function is highly valuable in various industries, from enhancing the umami flavor in fermented foods to serving as a therapeutic agent by depleting glutamine in cancer cells. Recent research highlights the diversity of these enzymes across different microbial sources. For example, studies have isolated promising glutaminases from bacteria such as *Streptomyces* species, with a recent investigation focusing on *Streptomyces canarius* for its efficiency against human cancer cell lines (Mostafa et al. 2021). Similarly, fungi from the genus *Aspergillus*, including *Aspergillus oryzae*, are widely recognized for their role in producing stable, extracellular glutaminases that are key to traditional food fermentations like soy sauce production (Jahanshahi and Tavakoli 2023).

Attempt is also made to explore the structural diversity and catalytic mechanisms of microbial glutaminases, highlighting their unique properties that make them suitable candidates for therapeutic development. This includes an in-depth analysis of their substrate specificity, kinetic parameters, and stability profiles under physiological conditions. Additionally, this chapter aims to delve into the various biotechnological approaches employed for the production and modification of these enzymes, including genetic engineering and immobilization techniques, to enhance their therapeutic properties and reduce immunogenicity in vivo (Tinoco et al. 2018; Janicki et al. 2023). The clinical viability of these enzymes is further enhanced by their low toxicity and unique substrate specificity, making them attractive candidates for therapeutic development across various medical conditions beyond oncology (Vanderstocken et al. 2022). For instance, some enzymes are engineered to activate prodrugs specifically at tumor sites, minimizing systemic toxicity and enhancing therapeutic indices (Vellard 2003).

Here, we have also discussed the challenges associated with the clinical translation of enzyme therapeutics, such as immunogenicity, short half-life, and delivery mechanisms, and discuss innovative strategies to overcome these hurdles. Moreover, the regulatory considerations and future perspectives for microbial glutaminases in clinical settings, emphasizing the ongoing research and development efforts aimed at broadening their therapeutic applicability, are also discussed. The increasing number of enzyme-based drugs approved by regulatory bodies underscores the growing recognition of their therapeutic efficacy and safety profiles, further solidifying their role in modern medicine (Vanderstocken et al. 2022). This includes a comprehensive review of the various microbial sources from which glutaminases have been isolated, alongside an examination of the genetic and biochemical modifications that can enhance their therapeutic properties (Santos-Beneit 2024). For example, enzymes can be engineered for improved stability, reduced immunogenicity, and enhanced catalytic activity, broadening their therapeutic window and patient applicability (Vellard 2003). Recent advancements in biotechnology have facilitated the development of more potent and specific enzyme therapeutics, leading to a significant increase in their commercial production and market demand, particularly for hydrolases which constitute nearly 75% of industrial enzymes (Vellard 2003; Fasim et al. 2020). The global enzyme market continues to expand rapidly, driven by innovations in bioreactor design and engineering interventions that enable more efficient and cost-effective production of these valuable biological catalysts (Tarafdar et al. 2021). This attempt is made to explore the diverse applications of microbial glutaminases beyond oncology, including their potential in treating metabolic disorders and infectious diseases. Specifically, the role of microbial glutaminases in modulating immune responses and degrading harmful by-products in metabolic pathways will be critically assessed. This includes an evaluation of their potential as antimicrobial agents through the disruption of microbial nitrogen metabolism and their utility in detoxification processes by breaking down toxic glutamine-derived compounds. Furthermore, the review will highlight recent discoveries of novel microbial glutaminases with unique enzymatic properties and discuss their potential for bioproduction using optimized bioprocesses (Wiltschi et al. 2020; Youssef et al. 2025). The expanding landscape of enzyme therapeutics suggests a promising future for microbial glutaminases, particularly as research continues to uncover their multifaceted roles and engineering strategies enhance their clinical viability (Vellard 2003).

The cloning of the glutaminase gene from a microbial source and expressing it in a different host organism has been used for higher yields and easier purification than from the native producer. A common host is *Escherichia coli*, which is favored for its rapid growth and well-understood genetics. For instance, the glutaminase gene from *Bacillus licheniformis*, an enzyme with potential in food flavoring, has been successfully expressed in *E. coli* to produce a recombinant enzyme with high efficiency (Sudhir et al. 2014). Similarly, researchers have utilized *E. coli* to express glutaminase genes from sources like *Streptomyces*, enabling the production and characterization of enzymes for potential use in anticancer therapies (Amobonye et al. 2019).

The purpose of this chapter is to offer a comprehensive review of microbial glutaminases. It aims to define the enzyme and discuss its key enzymatic properties, while also highlighting its widespread occurrence and diverse applications in the food and pharmaceutical industries. Furthermore, this chapter seeks to shed light on the structural characteristics and physicochemical properties of these enzymes, providing a foundation for future research and full exploration of their therapeutic and industrial potential.

8.2 Overview of Glutaminases

Glutaminase (EC 3.5.1.2), or L-glutaminase, is an amidase enzyme that catalyzes the hydrolysis of L-glutamine to glutamate and ammonia. Discovered in 1935 by Hans Krebs, glutaminase plays a key role in nitrogen metabolism in both prokaryotes and eukaryotes. Beyond its metabolic function, microbial glutaminase is commercially important as an oncolytic and antiretroviral agent, as well as a flavor enhancer in the food industry (Vijayan et al. 2017). Glutaminases play a foundational role in cellular metabolism, particularly in the process of nitrogen balance by catalyzing the hydrolysis of glutamine to produce glutamate and ammonia, a reaction essential for supporting numerous metabolic pathways and maintaining cellular nitrogen homeostasis. Besides this, glutaminases are key players in regulating critical cellular processes such as cell growth, differentiation, and immune responses. Due to this, glutaminases are explored for therapeutic interventions, especially in diseases characterized by uncontrolled cell proliferation or metabolic dysfunction. Glutaminases exist in various isoforms, each with distinct kinetic properties, regulatory mechanisms, and tissue-specific expression patterns. The structural and functional diversity of glutaminases is vital importance for understanding the various roles in different physiological and pathological conditions. For instance, certain glutaminase isoform may be highly active in tumor metabolism, while another may be predominantly involved in normal tissue glutamine utilization. The differential involvement allows for the development of highly selective therapeutic agents that can precisely modulate glutaminase activity in a targeted manner, minimizing off-target effects. Characterizing these isoforms is therefore essential for designing inhibitors or activators for therapeutic benefit (Spodenkiewicz et al. 2016). To harness the therapeutic potential of glutaminases, it is vital to employ proper methodologies for their isolation, purification, and characterization from microbial sources. Continuous efforts have been on understanding the structure–function of microbial glutaminases, increasing enzyme production using rational strategies, characterize glutaminase from extreme sources, and determine their enzymatic properties and substrate specificities critical for various potential applications (Nandakumar et al., 2003; Singh et al., 2013; Farahat et al. 2020; Amobonye et al., 2022); Sanchez-Moguel et al. 2023; Zhou et al. 2024; Wu et al., 2025). The therapeutic utility of microbial glutaminases extends beyond oncology to include infectious diseases and neurological disorders. An in-depth understanding of their properties informs the rational design of enzyme therapeutics, leading to the development of highly specific and potent inhibitors or activators (Noshahri and Rudat 2024).

8.3 Microbial Sources of Glutaminases

This section delves into the distinctive features of glutaminases derived from microbial sources, highlighting their unique biochemical characteristics, stability profiles, and production advantages compared to mammalian counterparts. Microbial glutaminases often exhibit enhanced thermal stability and broader pH optima, making them particularly suitable for various industrial and therapeutic applications. These enzymes are also amenable to large-scale production through fermentation processes, offering a cost-effective and scalable alternative to glutaminases sourced from animal tissues or recombinant mammalian systems (Vanderstocken et al.

2022). This section will detail the structural variations and catalytic mechanisms unique to microbial glutaminases, emphasizing how these properties contribute to their robust performance under diverse conditions. Furthermore, the genetic tractability of many microbial species allows for significant engineering modifications to optimize glutaminase activity, substrate specificity, and resistance to proteolysis, thereby expanding their utility in biotechnological and clinical settings (Vanderstocken et al. 2022). Various microorganisms, including bacteria, fungi, and yeasts, serve as prolific sources of glutaminases, with each organism potentially yielding enzymes with distinct properties (n.d.). For example, L-glutaminases derived from *Lactobacillus gasseri* have demonstrated antioxidant activity, suggesting their potential in nutraceutical and therapeutic applications (Hasson et al. 2021). Moreover, the production of microbial glutaminases can be further optimized through advanced genetic engineering techniques and bioprocess intensification, enabling tailored enzyme properties for specific applications. These advancements allow for the development of glutaminases with improved catalytic efficiency, altered substrate specificity, and enhanced stability, crucial for their successful translation into clinical and industrial products. The exploration of microbial glutaminases also extends to their potential as therapeutic agents, with specific attention given to their roles in cancer treatment, where they can deplete glutamine essential for tumor proliferation. Beyond oncology, microbial enzymes are increasingly recognized for their broader therapeutic utility, including applications in infectious disease control, where bacteriophage-derived enzymes can degrade bacterial cell walls, and in respiratory conditions, where certain enzymes can aid mucus clearance (Vanderstocken et al. 2022; Vellard 2003). This adaptability makes microbial glutaminases a focal point for developing novel biocatalysts and therapeutic agents with diverse applications (Boukid et al. 2023). Their utility in modulating gut microbiota, which plays a significant role in various physiological processes, is also being explored, positioning microbial glutaminases as promising candidates for novel therapeutic strategies. The scalable and cost-effective production of microbial enzymes further enhances their appeal for industrial and therapeutic applications, aligning with the growing demand for sustainable biotechnological solutions (Okpara 2022; Elazzazy et al. 2025). The ability to produce these enzymes economically and in large quantities positions microbial glutaminases as prime candidates for various biomedical applications, particularly in enzyme replacement therapies and as targeted anticancer agents (Vanderstocken et al. 2022). This highlights the critical role of fermentation and other bioprocess strategies in achieving economically viable production scales for these valuable enzymes (Tamano 2014). This emphasis on optimized production methods is particularly pertinent given the challenges associated with the stability and immunogenicity of enzyme therapeutics (Vanderstocken et al. 2022). Maintaining functional ability and addressing potential immunogenic responses are significant challenges in the development of enzyme therapeutics (Vanderstocken et al. 2022). Addressing these hurdles necessitates innovative approaches in protein engineering and formulation to enhance enzyme half-life and reduce adverse immune reactions (Vanderstocken et al. 2022). Furthermore, the burgeoning field of protein engineering offers promising avenues to enhance microbial glutaminases' therapeutic profiles by improving their pharmacokinetic properties and mitigating immunogenicity (Rettenbacher et al. 2021). This can involve strategies such as PEGylation, which has been shown to improve the half-life and reduce the immunogenicity of therapeutic enzymes (Vellard 2003), or directed evolution to select for variants with enhanced stability and reduced antigenicity. Beyond these modifications, the exploration of novel microbial strains with intrinsically superior glutaminase characteristics also presents a vital research direction, potentially yielding enzymes with naturally optimized therapeutic attributes. Such advancements are critical for translating promising enzyme candidates into effective clinical interventions, ultimately expanding the therapeutic repertoire available for complex diseases. The development and manufacturing of biopharmaceuticals, including microbial enzymes like glutaminases, has gained significant importance due to their growing demand in treating various conditions, particularly cancerous, immunological, and infectious diseases (Reger et al. 2023). The traditional methods for developing such recombinant proteins are often time-consuming, necessitating several years to progress from laboratory research to industrial production (Sun et al. 2023). However, recent advancements in cell culture bioprocessing, including the adoption of continuous processing and enhanced in-line sensing, aim to accelerate this development timeline and improve capital efficiency in therapeutic protein manufacturing (O'Brien and Hu 2020). This accelerated development is crucial for bringing novel enzyme-based therapies to market more rapidly, especially given the significant growth predicted for the global enzyme therapeutics market (Vanderstocken et al. 2022). The expanding therapeutic landscape for enzymes, with approximately 30 new enzyme drugs approved by the European Medicines Agency between 2000 and 2020, underscores a paradigm shift toward enzyme-based interventions for conditions previously lacking effective treatments, notably immune disorders and metabolic diseases (Vanderstocken et al. 2022; Vellard 2003). This expansion highlights the increasing recognition of enzymes as highly specific and functional therapeutic agents, capable of addressing complex biological pathways with minimal off-target effects (Vanderstocken et al. 2022). The successful development of biopharmaceuticals relies heavily on efficient expression systems, with mammalian cells, particularly Chinese hamster ovary cells, being the most frequently utilized due to their capacity for high-quality protein production and human-like post-translational modifications (Fischer et al. 2015; O'Flaherty et al. 2020). This ability to perform proper folding and glycosylation is critical for the biological activity and stability of complex recombinant proteins, making mammalian cell lines indispensable for the large-scale manufacturing of many therapeutic enzymes (Li et al. 2022; Butler and Meneses-Acosta 2012). However, the intricate nature of mammalian cell culture also presents challenges, including high production costs and susceptibility to viral contamination, driving continued research into alternative expression systems and purification strategies (Ōmasa et al. 2010). Consequently, advancements in microbial expression systems and transgenic animal models are being explored as viable alternatives to address these limitations, offering potential for more cost-effective and scalable production of therapeutic proteins (Jaramillo et al. 2011). Furthermore, the integration of advanced bioinformatics and synthetic biology approaches is poised to revolutionize the design and optimization of microbial glutaminases, enabling the creation of novel variants with tailored catalytic activities and improved therapeutic indices. Such interdisciplinary efforts are essential for fully realizing the potential of glutaminases in biomedicine, paving the way for targeted and highly effective treatments.

8.3.1 Bacterial Glutaminases

Numerous bacterial species, including those from the genera *Escherichia*, *Bacillus*, and *Pseudomonas*, are known to produce glutaminases with distinct biochemical properties and regulatory mechanisms. These microbial glutaminases often exhibit variations in substrate specificity, pH optima, and thermal stability, making them attractive candidates for diverse biotechnological applications (Yvon and Rijnen 2001). Their varied characteristics stem from structural adaptations that enable them to thrive in disparate environmental niches, offering valuable insights into enzyme evolution and function. For instance, some bacterial glutaminases display broad substrate specificity, acting on various glutamine derivatives, while others exhibit a high degree of selectivity for L-glutamine, reflecting their specific physiological roles in the host organism. Furthermore, the synthesis of these bacterial glutaminases is often tightly regulated by environmental factors, such as nutrient availability and pH, demonstrating sophisticated control mechanisms to optimize their metabolic contributions (Yvon and Rijnen 2001). The metabolic versatility of these enzymes, particularly their involvement in amino acid catabolism and nitrogen assimilation, underscores their importance for microbial survival and adaptation in diverse ecological niches (Torres et al. 2023).

Actinomycetes represent a significant source of glutaminases, often exhibiting properties distinct from those found in bacteria and fungi, making them promising candidates for industrial and therapeutic applications. These microbes are renowned for producing a wide array of secondary metabolites, including enzymes, with diverse bioactivities (Selim et al. 2021). Glutaminases derived from actinomycetes are particularly notable for their potential in enzyme therapy and bioremediation due to their robust catalytic activity under various conditions (Vanderstocken et al. 2022). Moreover, their unique metabolic pathways often lead to the production of glutaminases with novel enzymatic properties, such as enhanced stability or modified substrate specificities, which are highly desirable for biotechnological processes. The unique enzyme characteristics of actinomycete glutaminases, such as their resistance to feedback inhibition, could be harnessed for optimized therapeutic applications, including the development of novel anticancer agents or more efficient enzyme replacement therapies (Kloosterman et al. 2003).

8.3.2 Fungal Glutaminases

Fungal glutaminases, conversely, are pivotal in amino acid biosynthesis and nitrogen cycling within their ecosystems, with some species, particularly human pathogenic fungi, presenting unique glutaminase isoforms that are indispensable for their virulence and survival (Jastrzębowska and Gabriel 2014). These fungal enzymes often exhibit distinct biochemical properties, such as differing pH optima and substrate affinities, which distinguish them from their bacterial counterparts and highlight their specialized roles in fungal metabolism and pathogenesis (Yvon and Rijnen 2001).

8.4 Structure–Function Analysis of Microbial Glutaminases

The primary structure, defined by the linear sequence of amino acids, is highly conserved in key functional regions, suggesting evolutionary pressure for maintaining catalytic competence, but exhibits variability in less critical regions, contributing to adaptation to diverse ecological niches. Such sequence analysis often reveals conserved motifs, like the glutaminase signature sequence, which are directly involved in substrate binding or catalysis, thereby providing insights into the enzyme's mechanism of action. Moreover, comparative genomics and phylogenetics of microbial glutaminases can illuminate evolutionary relationships and identify novel enzymes with unique properties, potentially expanding the repertoire of therapeutic targets. Furthermore, analysis of the primary sequence can identify potential post-translational modification sites, which may modulate enzyme activity, stability, or localization within the microbial cell. The diversity in primary sequences also influences the enzyme's solubility and aggregation propensity, factors that are critically important for large-scale production and therapeutic formulation. Computational approaches, such as molecular dynamics simulations and machine learning algorithms, are increasingly being employed to predict the impact of specific amino acid substitutions on glutaminase structure and function, facilitating directed evolution strategies. These predictive models, combined with high-throughput screening methods, allow for the identification of advantageous mutations that enhance catalytic efficiency or alter substrate specificity, accelerating the development of superior glutaminase variants for clinical applications.

Microbial glutaminases are a diverse group of enzymes, each possessing unique structural features that dictate their substrate specificity, catalytic mechanism, and stability. These enzymes typically fall into the hydrolase class and catalyze the deamidation of glutamine to glutamate and ammonia. Their classification is primarily based on their molecular weight, subunit composition, and phylogenetic relationships. Further characterization of these structural attributes is crucial for understanding the functional differences among various glutaminases and for guiding rational design efforts aimed at optimizing their therapeutic efficacy (Li et al. 2022). Many microbial glutaminases exist as oligomers, commonly dimers or tetramers, with each subunit contributing to the overall active site architecture and structural integrity, which directly influences their catalytic efficiency and resistance to denaturation. The variability in their primary amino acid sequences leads to diverse tertiary and quaternary structures, which in turn affect their interaction with substrates and inhibitors. This structural diversity also underpins their adaptability to various physiological conditions, including pH and temperature extremes, making them attractive candidates for biotechnological applications. For instance, some microbial glutaminases exhibit a high degree of thermal stability, making them suitable for industrial processes requiring elevated temperatures, while others show enhanced activity at physiological pH, which is advantageous for therapeutic applications. Elucidating the intricate relationship between structural motifs and functional characteristics remains a paramount objective in glutaminase research, facilitating the engineering of enzymes with enhanced therapeutic profiles (Janicki et al. 2023). Notably, the glutamine:fructose-6-phosphate amidotransferase, a critical enzyme in the hexosamine biosynthetic pathway, highlights how subtle structural variations can lead to distinct functional roles, thus serving as a key regulator in metabolic diseases such as type 2 diabetes (Nakaishi et al. 2008). The catalytic mechanisms of microbial glutaminases typically involve a nucleophilic attack on the glutamine amide bond, often facilitated by a conserved catalytic triad or dyad, leading to the release of ammonia and glutamate. This process is highly regulated and can be influenced by allosteric effectors or post-translational modifications, which fine-tune the enzyme's activity in response to cellular needs. Understanding these mechanistic details is vital for the rational design of glutaminase inhibitors or activators, which could serve as therapeutic agents for a variety of conditions. Investigations into the structural variations of microbial glutaminases reveal distinct active site architectures and substrate-binding pockets that contribute to their specificities and catalytic efficiencies (Zhang et al. 2019).

The three-dimensional architecture of microbial glutaminases comprises distinct structural domains that contribute synergistically to their overall function, often involving an N-terminal regulatory domain and a C-terminal catalytic domain. These domains may function independently or cooperatively, allowing for precise control over enzymatic activity through allosteric regulation or domain–domain interactions. The presence of multiple domains with differing structural and functional properties has been observed in various enzymes, similar to how streptokinase possesses distinct domains that contribute to its diverse functions (Kunamneni et al. 2007). The intricate interplay between these domains is critical for maintaining the structural integrity and optimal catalytic activity of the glutaminase, ensuring its proper function within the cellular environment. The active site of glutaminases, often located at the interface of these domains, exhibits a remarkable degree of structural flexibility, which is crucial for accommodating various substrates and intermediates throughout the catalytic cycle (Weng et al. 2011). This dynamic nature facilitates induced fit mechanisms, where substrate binding triggers conformational changes that optimize the active site for catalysis, ensuring high specificity and efficiency. This adaptability extends to the potential for therapeutic intervention, as targeting these flexible regions could lead to the development of highly specific inhibitors that block the enzyme's activity. Moreover, advancements in computational modeling and machine learning have significantly improved the prediction of site-specific mutational effects on protein stability and function, offering new avenues for rational design of glutaminase variants with enhanced therapeutic profiles (Totaro et al. 2024). The incorporation of non-canonical amino acids into glutaminases through genetic code expansion techniques presents a novel strategy for further enhancing their stability, specificity, and catalytic efficiency, enabling the biosynthesis of custom-tailored enzymes for specific therapeutic applications (Chen et al. 2023). The genetically engineered enzymes, potentially endowed with bio-orthogonal labels, could serve as novel prodrugs or as tools for advanced bioimaging and kinetic studies, thereby broadening the scope of glutaminase-based therapies (Zou et al. 2018; Goettig et al. 2023).

The exploration of orthogonal active-site labels further enables detailed structural and mechanistic analyses of glutaminases, providing unprecedented insights into their catalytic machinery and inhibitory mechanisms (Jain et al. 2021). Such detailed structural and functional insights are paramount for the rational design of novel glutaminase inhibitors or activators, which could be critical for therapeutic interventions in various diseases, including cancer and neurological disorders (Rezhdo et al. 2019). Given that glutaminase is the primary glutamate-producing enzyme in the brain, understanding its precise regulation and diverse isoforms is critical for developing interventions targeting excitotoxic damage in neurological conditions (Márquez et al. 2009; Rosa et al. 2009). Understanding these structural and functional nuances is critical for developing targeted therapies that modulate glutaminase activity without inducing detrimental off-target effects. Indeed, subtle alterations in the active site's flexibility can significantly impact substrate promiscuity and catalytic efficiency, sometimes counterintuitively leading to increased activity and thermostability in promiscuous enzymes (Nutschel et al. 2021). The precise arrangement of amino acid residues within the active site pocket dictates the enzyme's binding affinity and catalytic rate, making it a prime target for drug design (Nováková et al. 2015). Specifically, detailed crystallographic and cryo-electron microscopy studies have provided atomic-level resolution of these active sites, and provided intricate hydrogen bonding networks and critical catalytic residues crucial for substrate recognition and conversion.

These structural insights are invaluable for the rational design of small molecule inhibitors or allosteric modulators that precisely fit within the active site or induce conformational changes that disrupt catalytic activity. For instance, the identification of key residues involved in substrate binding and catalysis can guide the development of transition state analogs or competitive inhibitors with high specificity and potency, minimizing off-target effects. Furthermore, advancements in computational chemistry, such as quantum mechanical/molecular mechanical (QM/MM) simulations, are now being employed to accurately model the electronic transitions and reaction mechanisms occurring within the glutaminase active site, offering unprecedented detail into the catalytic cycle. The detailed understanding can facilitate the rational design of novel glutaminase inhibitors with improved efficacy and reduced toxicity, thereby enhancing their therapeutic utility. The conversion of glutamine to glutamate, mediated by glutaminase, is a pivotal step in neuronal metabolism, underpinning both excitatory neurotransmission and energy production, especially under glucose-deprived conditions (Bheemanapally et al. 2020).

8.5 Biological Role of Glutaminases

Glutaminase, a critical enzyme in intermediary metabolism, catalyzes the hydrolytic deamidation of L-glutamine to L-glutamate and ammonia. This reaction is pivotal for various cellular processes, including amino acid catabolism, gluconeogenesis, and the biosynthesis of neurotransmitters in the central nervous system (Schousboe et al. 2013). Specifically, glutaminase activity is crucial for maintaining the glutamate–glutamine cycle, which is fundamental for excitatory neurotransmission (Bröer and Brookes 2001). Dysregulation of this enzymatic activity has been implicated in numerous pathological conditions, ranging from neurodegenerative disorders to cancer, underscoring its therapeutic relevance. In neural tissues, glutaminase isoforms, particularly phosphate-activated glutaminase, play a crucial role in the synthesis of the excitatory neurotransmitter glutamate, thereby influencing synaptic plasticity and cognitive functions (Peñalver et al. 2017). The precise control of glutaminase activity is therefore essential for neural health, as perturbations can lead to imbalances in glutamate levels, contributing to excitotoxicity or impaired neurotransmission (Kölker 2018; Scimemi 2019). Furthermore, the enzyme's role in supplying glutamate to the tricarboxylic acid cycle directly impacts cellular energy metabolism, particularly within astrocytes which play a significant role in maintaining brain glucose homeostasis (Chen et al. 2025; Schousboe et al. 2013). Astrocytes also play a critical role in regulating glutamatergic activity, influencing synaptic function, and mediating the uptake and metabolism of glutamate to prevent excitotoxicity and maintain normal brain function (Hertz and Zielke 2004). This metabolic interdependence underscores the complex homeostatic mechanisms regulating glutamate and glutamine levels within the brain, critical for both neuronal signaling and overall cerebral metabolic stability (Hayashi 2018). Beyond its role in neurotransmission, glutamate serves as a precursor for other essential molecules, such as GABA and glutathione, further emphasizing the multifaceted importance of glutaminase in maintaining cellular integrity and function (Niciu et al. 2011). Additionally, glutaminase activity contributes to ammonia detoxification, particularly in the liver and kidneys, where it facilitates the excretion of excess nitrogen. Moreover, the diverse physiological roles of glutaminase highlight its significance as a therapeutic target in various diseases where aberrant glutamine metabolism is a key driver of pathophysiology. Given its pervasive involvement in metabolic pathways critical for cellular proliferation and neurological function, understanding the structural nuances and regulatory mechanisms of microbial glutaminases offers promising avenues for targeted therapeutic interventions (Abdallah et al. 2014).

The conversion of glutamine to glutamate, mediated by glutaminase, is a pivotal step in neuronal metabolism, underpinning both excitatory neurotransmission and energy production, especially under glucose-deprived conditions (Bheemanapally et al. 2020). This metabolic pathway is integral to the glutamate–glutamine cycle, where glutamine serves as a critical precursor for the synthesis of neurotransmitter glutamate in neurons and its subsequent recycling in astrocytes (Zhang et al. 2024). A significant portion of astrocytic glutamine is transported to neurons, where phosphate-activated glutaminase deaminates it to glutamate, a reaction crucial for maintaining the excitatory neurotransmitter pool (Albrecht 2006). This process is tightly regulated to ensure proper neuronal function, as dysregulation of glutaminase activity can lead to excitotoxicity, a condition implicated in neurodegenerative diseases (Schousboe et al. 2014). Furthermore, the meticulous regulation of this glutamate–glutamine cycle, involving the rapid uptake of glutamate by astrocytes and its conversion to glutamine, prevents excessive neurotransmitter accumulation in the synaptic cleft, thereby safeguarding neuronal health (Pardo et al. 2013). Indeed, the enzymatic activity of glutaminase in neurons is essential for the synthesis of neurotransmitter glutamate, especially during development (Thomas et al. 1989). This metabolic interplay highlights the critical role of glutaminase in brain nitrogen trafficking and ammonia homeostasis, as it facilitates the conversion of glutamine, a crucial vehicle for ammonia detoxification, into glutamate (Bak et al. 2006; Schousboe et al. 2014). Given that ammonium is released during the glutaminase reaction, its transport from the neuronal mitochondria to the astrocytic cytosol is critical for continuous glutamine production within the glutamate–glutamine cycle (Hertz 2013). This delicate balance underscores the necessity for precise regulation of glutaminase activity, as its dysregulation can perturb the intricate metabolic and neurochemical equilibrium essential for brain function (Shen 2013). The interplay between astrocytic glutamine synthesis and neuronal glutaminase activity thus represents a critical nexus for maintaining synaptic efficacy and preventing neurotoxic glutamate accumulation (Daikhin and Yudkoff 2000). Moreover, the dynamic regulation of intracellular pH within astrocytes and neurons is intricately linked to glutamine/glutamate metabolism, influencing the efficiency of ammonia detoxification and energy supply to neurons (Nissim 2012). The glial–neuronal metabolic coupling, mediated by glutaminase, is thus central to sustaining neuronal excitability and protecting against excitotoxic damage by orchestrating efficient glutamate recycling (Albrecht et al. 2010; Shen 2013). This intricate metabolic symphony highlights the brain's remarkable capacity for maintaining energy homeostasis and neurotransmitter balance through precise enzymatic regulation and intercellular metabolic channeling (Waagepetersen et al. 2005; Coulter and Eid 2012).

8.6 Therapeutic Applications of Glutaminases

Glutaminases have garnered considerable interest in the pharmaceutical industry due to their involvement in critical metabolic pathways, particularly in cancer and infectious diseases. The ability of these enzymes to deplete glutamine, an amino acid crucial for the proliferation of many cancer cells, positions them as promising candidates for antineoplastic therapies, mirroring the success of asparaginases in acute lymphoblastic leukemia (Pokrovsky et al. 2022). This approach leverages the metabolic vulnerabilities of rapidly dividing cells, which often exhibit an increased reliance on glutamine for energy production and biosynthesis (Vellard 2003). This dependency arises because malignant cells frequently upregulate glutamine metabolism to support their rapid growth and survival, making glutaminase-mediated glutamine depletion an attractive therapeutic strategy (Endicott et al. 2021). Furthermore, glutaminases from microbial sources, including bacteria and fungi, have shown therapeutic potential in various medical conditions beyond oncology, such as hyperuricemia and certain genetic disorders (Vellard 2003). Their broad-spectrum utility underscores the therapeutic versatility of glutaminases, warranting further investigation into their application in diverse clinical settings. One significant area of application for microbial glutaminases is their role as antineoplastic agents, primarily due to their capacity to deplete glutamine, an essential nutrient for the unchecked proliferation of numerous cancer cells. This strategy exploits the Warburg effect, a metabolic hallmark of many cancers, where cells primarily rely on aerobic glycolysis and glutaminolysis to sustain their rapid growth and biosynthetic demands (Vellard 2003).

Inhibition of glutamine metabolism by glutaminases effectively starves these highly proliferative cells, leading to cell cycle arrest and apoptosis (Wilder et al. 2021). Consequently, researchers have focused on developing microbial glutaminases with enhanced stability, reduced immunogenicity, and targeted delivery mechanisms to improve their therapeutic efficacy in cancer treatment. In addition to their role in oncology, glutaminases have shown potential in treating various other diseases, including infectious diseases, where modulating glutamine availability can impact pathogen viability and host immune responses (Vanderstocken et al. 2022). For instance, the dependency of rapidly proliferating cells, including many cancer cells, on glutamine as a crucial nutrient for growth and proliferation can be exploited therapeutically, as these cells often exhibit altered nitrogen metabolism, making them particularly vulnerable to glutamine deprivation (Dang 2011; Kurmi and Haigis 2020). Targeting glutamine metabolism with microbial glutaminases represents a promising avenue for developing novel therapeutics against such diseases (Sborov et al. 2014). Beyond cancer, glutaminases have demonstrated efficacy in addressing hyperuricemia by modulating uric acid levels, and in enzyme replacement therapies for metabolic disorders (Vellard 2003). Furthermore, the unique substrate specificity of various microbial glutaminases allows for highly targeted interventions, minimizing off-target effects and enhancing their therapeutic index (Vanderstocken et al. 2022). Enzymes, in general, are widely recognized for their therapeutic applications, ranging from cancer treatment to metabolic disorders, leveraging their catalytic properties to elicit potent biological effects (Vanderstocken et al. 2022). This expanding recognition of enzymes' therapeutic utility is evidenced by the exponential growth in enzyme-based drug approvals, particularly in the last two decades (Vanderstocken et al. 2022). The success of enzyme therapy, such as Adagen for severe combined immunodeficiency, underscores the potential of enzymatic interventions for inherited diseases, often enhanced through modifications like PEGylation to improve half-life and reduce immunogenicity (Vellard 2003).

8.7 Challenges and Future Directions

The successful translation of microbial glutaminases and their inhibitors from laboratory to clinic faces several challenges, including optimizing enzyme stability, reducing immunogenicity, and ensuring targeted delivery to mitigate off-target effects. Further research is needed to overcome these hurdles and fully harness the therapeutic potential of these promising biomolecules. A significant area of ongoing investigation involves engineering glutaminases with enhanced catalytic efficiency and reduced immunogenicity through rational design and directed evolution, thereby improving their pharmacokinetic profiles and therapeutic index. Additionally, developing more precise drug delivery systems, such as nanoparticles or antibody–drug conjugates, is crucial to ensure selective accumulation in target tissues and minimize systemic toxicity. Moreover, exploring the synergistic effects of glutaminases with existing chemotherapies or immunotherapies holds promise for developing more effective combination treatments for various malignancies (Vanderstocken et al. 2022). Beyond oncology, further research into the application of microbial glutaminases in diverse medical conditions, such as inflammatory diseases and antibiotic-resistant infections, is warranted (Vanderstocken et al. 2022). This expanding therapeutic landscape for microbial glutaminases necessitates comprehensive preclinical and clinical studies to validate their efficacy and safety across various indications (Vellard 2003). The complex interplay between host metabolism and microbial glutaminases also presents avenues for developing novel diagnostic tools and personalized treatment strategies. Consequently, the development of predictive biomarkers that identify patients most likely to respond to glutaminase-based therapies is paramount for realizing precision medicine approaches. Furthermore, the identification and characterization of novel microbial glutaminases with distinct enzymatic properties and broader substrate specificities could unlock new therapeutic modalities, expanding their utility beyond current applications (Youssef et al. 2025). The ongoing exploration of microbial diversity and advancements in synthetic biology are poised to yield glutaminases with tailored characteristics, opening new frontiers for therapeutic intervention in an array of pathologies. However, challenges such as co-factor regeneration, substrate, and product inhibition remain significant limitations to widespread industrial enzyme application, necessitating innovative strategies to enhance their therapeutic utility (Noshahri and Rudat 2024). Indeed, improving enzyme stability, particularly at temperatures exceeding the glass transition of typical plastics, is crucial for broadening their industrial applicability to various hydrolytic enzymes and polymer classes (Huaman et al. 2024).

This is particularly relevant for the development of robust glutaminases capable of maintaining activity under diverse physiological conditions, enhancing their therapeutic viability. Furthermore, the exploration of genetic engineering and directed evolution techniques can significantly enhance the thermostability and catalytic efficiency of microbial glutaminases, making them more suitable for pharmaceutical applications (Vanderstocken et al. 2022). This proactive approach to enzyme design is critical for overcoming existing translational barriers and realizing the full therapeutic promise of glutaminases in a clinical setting (Wang et al. 2025; Vellard 2003). Another crucial aspect involves the development of bio-based technologies to ensure that glutaminase production is sustainable and environmentally sound, aligning with global efforts to minimize environmental impact (Orlando et al. 2023).

8.8 Conclusion

Glutaminase (GLS) has emerged as a pivotal enzyme in cellular metabolism, linking glutamine utilization to energy production, biosynthesis, and redox homeostasis. Over the past decade, a surge of biochemical, structural, and translational research has illuminated its central role in both health and disease. The expansive exploration into microbial glutaminases and their intricate mechanisms has illuminated their profound therapeutic potential across diverse medical fields. This includes their pivotal role in oncology, where glutamine depletion offers a potent strategy against various malignancies, and their broader utility in metabolic disorders and infectious diseases. Future research endeavors are poised to unravel the full spectrum of their biomedical applications, from precise enzyme replacement therapies to sophisticated immunomodulatory agents, paving the way for innovative clinical interventions (Vanderstocken et al. 2022). Further advancements in genetic engineering and synthetic biology are expected to unlock previously uncharacterized microbial products, offering novel therapeutic enzymes and enhancing existing ones for medical applications (Santos-Beneit 2024). This will allow for the development of highly specific and efficient microbial glutaminases, potentially leading to more effective and safer therapeutic agents (Giurini et al. 2024). The integration of computational biology and high-throughput screening methodologies will further accelerate the discovery and optimization of glutaminase variants with tailored enzymatic properties for specific therapeutic targets. This includes exploring the potential of non-canonical amino acids as scaffolds for novel enzyme therapeutics, broadening the scope of enzyme-based interventions (Martínez-Rodríguez et al. 2020). Specifically, novel glutaminase engineering approaches, such as directed evolution and rational design, hold promise for generating enzymes with improved catalytic efficiency, stability under physiological conditions, and reduced immunogenicity, crucial for chronic therapeutic administration (Youssef et al. 2025).

References

Abdallah CG, Jiang L, Feyter HMD, Fasula M, Krystal JH, Rothman DL, Mason GF, Sanacora G (2014) Glutamate metabolism in major depressive disorder. Am J Psychiatry 171(12):1320. https://doi.org/10.1176/appi.ajp.2014.14010067

Albrecht J (2006) Glutamine in the central nervous system: function and dysfunction Front Biosci 12(1):332. https://doi.org/10.2741/2067

Albrecht J, Sidoryk-Węgrzynowicz M, Zielińska M, Aschner M (2010) Roles of glutamine in neurotransmission. Neuron Glia Biol 6(4):263. Cambridge University Press. https://doi.org/10.1017/s1740925x11000093

Al-Thawadi S (2011) Fungal glutaminase and its role in cancer therapy: a review. J Biomed Biotechnol 2011:1–10

Amobonye A, Singh S, Pillai S (2019) Recent advances in microbial glutaminase production and applications—a concise review. J Appl Microbiol 127(5):1277–1289

Amobonye A, Singh S, Mukherjee K, Jobichen C, Qureshi IA, Pillai S (2022) Structural and functional insights into fungal glutaminase using a computational approach. Process Biochem 117:76–89

Bak LK, Schousboe A, Waagepetersen HS (2006) The glutamate/GABA-glutamine cycle: aspects of transport, neurotransmitter homeostasis and ammonia transfer. J Neurochem. 98(3):641–653. https://doi.org/10.1111/j.1471-4159.2006.03913.x. Epub 2006 Jun 19. PMID: 16787421

Bheemanapally K, Ibrahim MMH, Briski KP (2020) Optimization of ultra-high-performance liquid chromatography-electrospray ionization-mass spectrometry detection of glutamine-FMOC ad-hoc derivative by central composite design. Sci Rep 10(1). https://doi.org/10.1038/s41598-020-64099-w

Boukid F, Ganeshan S, Wang Y, Tülbek M, Nickerson MT (2023) Bioengineered enzymes and precision fermentation in the food industry [Review of bioengineered enzymes and precision fermentation in the food industry]. Int J Mol Sci 24(12):10156. Multidisciplinary Digital Publishing Institute. https://doi.org/10.3390/ijms241210156

Bröer S, Brookes N (2001) Transfer of glutamine between astrocytes and neurons. J Neurochem 77(3):705–719. https://doi.org/10.1046/j.1471-4159.2001.00322.x. PMID: 11331400.

Butler M, Meneses-Acosta A (2012) Recent advances in technology supporting biopharmaceutical production from mammalian cells. Appl Microbiol Biotechnol 96(4):885–894. https://doi.org/10.1007/s00253-012-4451-z. Epub 2012 Oct 5. PMID: 23053101; PMCID: PMC7080107

Chen L, Xin X, Zhang Y, Li S, Zhao X, Li S, Xu Z (2023) Advances in biosynthesis of non-canonical amino acids (ncAAs) and the methods of ncAAs incorporation into proteins. Molecules 28(18):6745. https://doi.org/10.3390/molecules28186745. PMID: 37764520; PMCID: PMC10534643.

Chen Q, Wang Y, Li R, Bai Q, Zhao Y (2025) The induced-fit and catalytic mechanisms of human G6PC1. Cell Discov 11(1). https://doi.org/10.1038/s41421-025-00814-z

Coulter DA, Eid T (2012) Astrocytic regulation of glutamate homeostasis in epilepsy. Glia 60(8):1215–1226. https://doi.org/10.1002/glia.22341. Epub 2012 May 16. PMID: 22592998; PMCID: PMC3375386

Daikhin Y, Yudkoff M (2000) Compartmentation of brain glutamate metabolism in neurons and glia [Review of *compartmentation of brain glutamate metabolism in neurons and glia*]. J Nutr 130(4). Elsevier BV. https://doi.org/10.1093/jn/130.4.1026s

Dang CV (2011) Therapeutic targeting of Myc-reprogrammed cancer cell metabolism [Review of *therapeutic targeting of Myc-reprogrammed cancer cell metabolism*]. Cold Spring Harb Symp Quant Biol 76:369. Cold Spring Harbor Laboratory Press. https://doi.org/10.1101/sqb.2011.76.011296

Elazzazy AM, Baeshen MN, Alasmi KM, Alqurashi SI, Desouky SE, Khattab SMR (2025) Where biology meets engineering: scaling up microbial nutraceuticals to bridge nutrition, therapeutics, and global impact [Review of where biology meets engineering: scaling up microbial nutraceuticals to bridge nutrition, therapeutics, and global impact]. Microorganisms 13(3):566. Multidisciplinary Digital Publishing Institute. https://doi.org/10.3390/microorganisms13030566

Endicott M, Jones M, Hull J (2021) Amino acid metabolism as a therapeutic target in cancer: a review. Amino Acids 53(8):1169–1179. https://doi.org/10.1007/s00726-021-03052-1. Epub 2021 Jul 22. PMID: 34292410; PMCID: PMC8325646.

Farahat MG, Amr D, Galal A (2020) Molecular cloning, structural modeling and characterization of a novel glutaminase-free L-asparaginase from Cobetia amphilecti AMI6. Int J Biol Macromol 143:685–695

Fasim A, More VS, More SS (2021) Large-scale production of enzymes for biotechnology uses. Curr Opin Biotechnol 69:68–76. https://doi.org/10.1016/j.copbio.2020.12.002. Epub 2020 Dec 31. PMID: 33388493

Fischer S, Handrick R, Otte K (2015) The art of CHO cell engineering: a comprehensive retrospect and future perspectives. Biotechnol Adv 33(8):1878–1896. https://doi.org/10.1016/j.biotechadv.2015.10.015. Epub 2015 Oct 31. PMID: 26523782

Giurini EF, Godla A, Gupta K (2024) Redefining bioactive small molecules from microbial metabolites as revolutionary anticancer agents. Cancer Gene Ther 31(2):187. https://doi.org/10.1038/s41417-023-00715-x

Goettig P, Koch NG, Budisa N (2023) Non-canonical amino acids in analyses of protease structure and function. Int J Mol Sci 24(18):14035. https://doi.org/10.3390/ijms241814035. PMID: 37762340; PMCID: PMC10531186

Hasson B, Mahdi L, Essa RH (2021) Evidence of antioxidant activity of novel L-Glutaminase purified from L. Gasseri BRLHM. J Appl Sci Nanotechnol 1(4):44. https://doi.org/10.53293/jasn.2021.3969.1064

Hayashi MK (2018) Structure-function relationship of transporters in the glutamate-glutamine cycle of the central nervous system. Int J Mol Sci 19(4):1177. https://doi.org/10.3390/ijms19041177. PMID: 29649168; PMCID: PMC5979278

Hertz L (2013) The glutamate–glutamine (GABA) cycle: importance of late postnatal development and potential reciprocal interactions between biosynthesis and degradation. Front Endocrinol 4. https://doi.org/10.3389/fendo.2013.00059

Hertz L, Zielke HR (2004) Astrocytic control of glutamatergic activity: astrocytes as stars of the show. Trends Neurosci 27(12):735–743. https://doi.org/10.1016/j.tins.2004.10.008. PMID: 15541514

Huaman SMM, Nicholson JH, Brogan APS (2024) A general route to retooling hydrolytic enzymes toward plastic degradation. Cell Rep Phys Sci 5(2):101783. https://doi.org/10.1016/j.xcrp.2024.101783

Jahanshahi M, Tavakoli H (2023) Fungal L-glutaminase: production, optimization, and purification with antimicrobial activities of L-glutaminase from novel isolate *Aspergillus tamarii* AUMC 10198 under solid-state fermentation. Biocatal Agric Biotechnol 47:102570

Jain N, Tamura K, Déjean G, Petegem FV, Brumer H (2021) Orthogonal active-site labels for mixed-linkage endo-β-Glucanases. ACS Chem Biol 16(10):1968. https://doi.org/10.1021/acschembio.1c00063

Janicki M, Ściuk A, Zieleziński A, Ruszkowski M, Ludwików A, Karłowski WM, Jaskólski M, Loch JI (2023) The effects of nature-inspired amino acid substitutions on structural and biochemical properties of the E. coli L-asparaginase EcAIII. Protein Sci 32(6). https://doi.org/10.1002/pro.4647

Jaramillo EH, Sensi A, Brandenberg O, Ghosh K, Sonnino A (2011) Biosafety resource book. Food and Agriculture Organization of the United Nations, Rome

Jastrzębowska K, Gabriel I (2015) Inhibitors of amino acids biosynthesis as antifungal agents. Amino Acids 47(2):227–249. https://doi.org/10.1007/s00726-014-1873-1. Epub 2014 Nov 20. PMID: 25408465; PMCID: PMC4302243

Kloosterman H, Hessels GI, Vrijbloed JW, Euverink GJW, Dijkhuizen L (2003) (De)regulation of key enzyme steps in the shikimate pathway and phenylalanine-specific pathway of the actinomycete Amycolatopsis methanolica. Microbiology 149(11):3321. https://doi.org/10.1099/mic.0.26494-0

Kölker S (2018) Metabolism of amino acid neurotransmitters: the synaptic disorder underlying inherited metabolic diseases. J Inherit Metab Dis 41(6):1055–1063. https://doi.org/10.1007/s10545-018-0201-4. Epub 2018 Jun 4. PMID: 29869166

Kunamneni A, Abdelghani TT, Ellaiah P (20077) Streptokinase--the drug of choice for thrombolytic therapy. J Thromb Thrombolysis 23(1):9–23. https://doi.org/10.1007/s11239-006-9011-x. PMID: 17111203.

Kurmi K, Haigis MC (2020) Nitrogen metabolism in cancer and immunity [Review of nitrogen metabolism in cancer and immunity]. Trends Cell Biol 30(5):408. Elsevier BV. https://doi.org/10.1016/j.tcb.2020.02.005

Li Z, Fan Z, Wang X-Y, Wang T (2022) Factors affecting the expression of recombinant protein and improvement strategies in Chinese hamster ovary cells [Review of factors affecting the

expression of recombinant protein and improvement strategies in Chinese hamster ovary cells]. Front Bioeng Biotechnol 10. Frontiers Media. https://doi.org/10.3389/fbioe.2022.880155
Márquez J, Tosina M, de la Rosa V, Segura JA, Alonso FJ, Matés JM, Campos-Sandoval JA (2009) New insights into brain glutaminases: beyond their role on glutamatergic transmission. Neurochem Int 55(1-3):64–70. https://doi.org/10.1016/j.neuint.2009.02.022. Epub 2009 Mar 9. PMID: 19428808
Martínez-Rodríguez S, Torres JM, Sánchez P, Ortega E (2020) Overview on multienzymatic cascades for the production of non-canonical α-amino acids. Front Bioeng Biotechnol 8:887. https://doi.org/10.3389/fbioe.2020.00887. PMID: 32850740; PMCID: PMC7431475
Moldovan O-L, Sandulea A, Lungu I-A, Gâz AŞ, Rusu A (2023) Identification of some glutamic acid derivatives with biological potential by computational methods. Molecules 28(10):4123. https://doi.org/10.3390/molecules28104123
Mostafa YS, Alamri SA, Alfaifi MY, Alrumman SA, Elbehairi SEI, Taha TH, Hashem M (2021) L-Glutaminase synthesis by marine *Halomonas meridiana* isolated from the Red Sea and its efficiency against colorectal cancer cell lines. Molecules 26(7):1963
Nakaishi Y, Bando M, Shimizu H, Watanabe K, Goto F, Tsuge H, Kondo K, Komatsu M (2008) Structural analysis of human glutamine:fructose-6-phosphate amidotransferase, a key regulator in type 2 diabetes. FEBS Lett 583(1):163. https://doi.org/10.1016/j.febslet.2008.11.041
Nandakumar R, Yoshimune K, Wakayama M, Moriguchi M (2003) Microbial glutaminase: biochemistry, molecular approaches and applications in the food industry. J Mol Catal B Enzym 23(2–6):87–100
Niciu MJ, Kelmendi B, Sanacora G (2012) Overview of glutamatergic neurotransmission in the nervous system. Pharmacol Biochem Behav 100(4):656–664. https://doi.org/10.1016/j.pbb.2011.08.008. Epub 2011 Aug 26. PMID: 21889952; PMCID: PMC3253893
Noshahri NG, Rudat J (2024) The promising role of amine transaminase cascades in the synthesis of non-canonical amino acids. Processes 12(11):2566. https://doi.org/10.3390/pr12112566
Nováková Z, Černý J, Choy CJ, Nedrow JR, Choi JK, Łubkowski J, Berkman CE, Bařinka C (2015) Design of composite inhibitors targeting glutamate carboxypeptidase II: the importance of effector functionalities. FEBS J 283(1):130. https://doi.org/10.1111/febs.13557
Nutschel C, Coscolín C, David B, Mulnaes D, Ferrer M, Jaeger K, Gohlke H (2021) Promiscuous Esterases counterintuitively are less flexible than specific ones. J Chem Inf Model 61(5):2383. https://doi.org/10.1021/acs.jcim.1c00152
O'Brien SA, Hu W (2020) Cell culture bioprocessing — the road taken and the path forward. Curr Opin Chem Eng 30:128. https://doi.org/10.1016/j.coche.2020.100663
O'Flaherty R, Bergin A, Flampouri E, Mota LM, Obaidi I, Quigley A, Xie Y, Butler M (2020) Mammalian cell culture for production of recombinant proteins: a review of the critical steps in their biomanufacturing [Review of mammalian cell culture for production of recombinant proteins: a review of the critical steps in their biomanufacturing]. Biotechnol Adv 43:107552. Elsevier BV. https://doi.org/10.1016/j.biotechadv.2020.107552
Okpara M (2022) Microbial enzymes and their applications in food industry: a mini-review. Adv Enzyme Res 10(1):23. https://doi.org/10.4236/aer.2022.101002
Ōmasa T, Onitsuka M, Kim W-D (2010) Cell engineering and cultivation of Chinese hamster ovary (CHO) cells [Review of cell engineering and cultivation of Chinese hamster ovary (CHO) cells]. Curr Pharm Biotechnol 11(3):233. Bentham Science Publishers. https://doi.org/10.2174/138920110791111960
Orlando M, Molla G, Castellani P, Pirillo V, Torretta V, Ferronato N (2023) Microbial enzyme biotechnology to reach plastic waste circularity: current status, problems and perspectives. Int J Mol Sci 24(4):3877. https://doi.org/10.3390/ijms24043877. PMID: 36835289; PMCID: PMC9967032
Pardo B, Contreras L, Satrústegui J (2013) De novo synthesis of glial glutamate and glutamine in young mice requires aspartate provided by the neuronal mitochondrial aspartate-glutamate carrier Aralar/AGC1. Front Endocrinol (Lausanne) 4:149. https://doi.org/10.3389/fendo.2013.00149. PMID: 24133485; PMCID: PMC3796713

Peñalver A, Campos-Sandoval JA, Blanco E, Cardona C, Castilla L, Martín-Rufián M, Estivill-Torrús G, Sánchez-Varo R, Alonso FJ, Pérez-Hernández M, Colado MI, Gutiérrez A, de Fonseca FR, Márquez J (2017) Glutaminase and MMP-9 downregulation in cortex and hippocampus of LPA1 receptor null mice correlate with altered dendritic spine plasticity. Front Mol Neurosci 10. https://doi.org/10.3389/fnmol.2017.00278

Pokrovsky VS, Qoura LA, Morozova EA, Bunik VI (2022) Predictive markers for efficiency of the amino-acid deprivation therapies in cancer. Front Med 9. https://doi.org/10.3389/fmed.2022.1035356

Reger LN, Saballus M, Kappes A, Kampmann M, Wijffels RH, Martens DE, Niemann J (2023) A novel hybrid bioprocess strategy addressing key challenges of advanced biomanufacturing. Front Bioeng Biotechnol 11. https://doi.org/10.3389/fbioe.2023.1211410

Rettenbacher LA, Arauzo-Aguilera K, Buscajoni L, Castillo-Corujo A, Ferrero-Bordera B, Kostopoulou A, Morán-Torres R, Núñez-Nepomuceno D, Öktem A, Palma A, Pisent B, Puricelli M, Schilling T, Tungekar AA, Walgraeve J, Humphreys DP, von der Haar T, Gasser B, Mattanovich D et al (2021) Microbial protein cell factories fight back? Trends Biotechnol 40(5):576. https://doi.org/10.1016/j.tibtech.2021.10.003

Rezhdo A, Islam M, Huang M, Van Deventer JA (2019) Future prospects for noncanonical amino acids in biological therapeutics. Curr Opin Biotechnol 60:168–178. https://doi.org/10.1016/j.copbio.2019.02.020. Epub 2019 Apr 8. PMID: 30974337; PMCID: PMC6783319

de la Rosa V, Campos-Sandoval JA, Martín-Rufián M, Cardona C, Matés JM, Segura JA, Alonso FJ, Márquez J (2009) A novel glutaminase isoform in mammalian tissues. Neurochem Int 55(1–3):76–84. https://doi.org/10.1016/j.neuint.2009.02.021. Epub 2009 Mar 9. PMID: 19428810

Sanchez-Moguel I, Costa-Silva TA, Pillaca-Pullo OS, Flores-Santos JC, Freire RKB, Carretero G et al (2023) Antarctic yeasts as a source of L-asparaginase: characterization of a glutaminase-activity free L-asparaginase from psychrotolerant yeast Leucosporidium scottii L115. Process Biochem 129:121–132

Santos-Beneit F (2024) What is the role of microbial biotechnology and genetic engineering in medicine? MicrobiologyOpen 13(2). https://doi.org/10.1002/mbo3.1406

Sborov DW, Haverkos BM, Harris PJ (2014) Investigational cancer drugs targeting cell metabolism in clinical development. Expert Opin Investig Drugs 24(1):79. https://doi.org/10.1517/13543784.2015.960077

Schousboe A, Bak LK, Waagepetersen HS (2013) Astrocytic control of biosynthesis and turnover of the neurotransmitters glutamate and GABA. Front Endocrinol 4. https://doi.org/10.3389/fendo.2013.00102

Schousboe A, Scafidi S, Bak LK, Waagepetersen HS, McKenna MC (2014) Glutamate metabolism in the brain focusing on astrocytes. In: Advances in neurobiology. Springer International Publishing, p 13. https://doi.org/10.1007/978-3-319-08894-5_2

Scimemi A (2019) The role of astrocytes in neurotransmitter uptake and brain metabolism, Springer series in computational neuroscience. Springer Nature, p 309. https://doi.org/10.1007/978-3-030-00817-8_12

Selim MSM, Abdelhamid SA, Mohamed SS (2021) Secondary metabolites and biodiversity of actinomycetes. J Genet Eng Biotechnol 19(1):72. https://doi.org/10.1186/s43141-021-00156-9. PMID: 33982192; PMCID: PMC8116480

Shen J (2013) Modeling the glutamate–glutamine neurotransmitter cycle. Front Neuroenerg 5. https://doi.org/10.3389/fnene.2013.00001

Singh P, Shera SS, Banik J, Banik RM (2013) Optimization of cultural conditions using response surface methodology versus artificial neural network and modeling of L-glutaminase production by Bacillus cereus MTCC 1305. Bioresour Technol 137:261–269

Spodenkiewicz M, Diez-Fernandez C, Rüfenacht V, Gemperle-Britschgi C, Häberle J (2016) Minireview on glutamine synthetase deficiency, an ultra-rare inborn error of amino acid biosynthesis. Biology (Basel) 5(4):40. https://doi.org/10.3390/biology5040040. PMID: 27775558; PMCID: PMC5192420

Sudhir AP, Dave BR, Prajapati AS, Panchal K, Patel D, Subramanian RB (2014) Characterization of a recombinant glutaminase-free L-asparaginase (ansA3) enzyme with high catalytic activity from Bacillus licheniformis. Appl Biochem Biotechnol 174(7):2504–2515

Sun M, Gao AX, Liu X, Yang Y, Ledesma-Amaro R, Bai Z (2023) High-throughput process development from gene cloning to protein production. Microb Cell Fact 22(1):182. https://doi.org/10.1186/s12934-023-02184-1. PMID: 37715258; PMCID: PMC10503041

Tamano K (2014) Enhancing microbial metabolite and enzyme production: current strategies and challenges. Front Microbiol 5. https://doi.org/10.3389/fmicb.2014.00718

Tarafdar A, Sirohi R, Gaur VK, Kumar S, Sharma P, Varjani S, Pandey HO, Sindhu R, Madhavan A, Rajasekharan R, Sim SJ (2021) Engineering interventions in enzyme production: lab to industrial scale. Bioresour Technol 326:124771. https://doi.org/10.1016/j.biortech.2021.124771. Epub 2021 Jan 30. Erratum in: Bioresour Technol. 2022 Oct;361:127770. https://doi.org/10.1016/j.biortech.2022.127770. PMID: 33550211

Thomas JW, Novelli A, Tao-Cheng J-H, Henneberry RC, Smith HH, Banner C (1989) Developmental induction of glutaminase in primary cultures of cerebellar granule cells. Mol Brain Res 6(1):47. https://doi.org/10.1016/0169-328x(89)90027-2

Tinoco A, Sárria MP, Loureiro A, Parpot P, Espiña B, Gomes AC, Cavaco-Paulo A, Ribeiro A (2018) BSA/ASN/Pol407 nanoparticles for acute lymphoblastic leukemia treatment. Biochem Eng J 141:80. https://doi.org/10.1016/j.bej.2018.10.006

Torres N, Tobón-Cornejo S, Velazquez-Villegas LA, Noriega LG, Alemán-Escondrillas G, Tovar AR. Amino Acid Catabolism: An Overlooked Area of Metabolism. Nutrients. 2023 Jul 29;15(15):3378. doi: 10.3390/nu15153378. PMID: 37571315; PMCID: PMC10421169.

Totaro MG, Vide U, Zausinger R, Winkler A, Oberdorfer G (2024) ESM-scan—a tool to guide amino acid substitutions. Protein Sci 33(12). https://doi.org/10.1002/pro.5221

Vanderstocken G, Woolf NL, Trigiante G, Jackson J, McGoldrick R (2022) Harnessing the potential of enzymes as inhaled therapeutics in respiratory tract diseases: a review of the literature. Biomedicines 10(6):1440. https://doi.org/10.3390/biomedicines10061440. PMID: 35740461; PMCID: PMC9220205

Vellard M (2003) The enzyme as drug: application of enzymes as pharmaceuticals. Curr Opin Biotechnol 14(4):444–450. https://doi.org/10.1016/s0958-1669(03)00092-2. PMID: 12943856.

Vijayan N, Swapna TS, Haridas M, Sabu A (2017) Therapeutic enzymes: l-glutaminase. In: Current developments in biotechnology and bioengineering. Elsevier, pp 233–248

Waagepetersen HS, Qu H, Sonnewald U, Shimamoto K, Schousboe A (2005) Role of glutamine and neuronal glutamate uptake in glutamate homeostasis and synthesis during vesicular release in cultured glutamatergic neurons. Neurochem Int 47:92. https://doi.org/10.1016/j.neuint.2005.04.012

Wang R, Zhuang J, Zhang Q, Wu W, Yu X, Zhang H, Xie Z (2025) Decoding the metabolic dialogue in the tumor microenvironment: from immune suppression to precision cancer therapies. Exp Hematol Oncol 14(1):99. https://doi.org/10.1186/s40164-025-00689-6. PMID: 40696413; PMCID: PMC12282028

Weng Y-Z, Chang DT-H, Huang Y, Lin C (2011) A study on the flexibility of enzyme active sites. BMC Bioinform 12. https://doi.org/10.1186/1471-2105-12-s1-s32

Wilder CS, Chen Z, DiGiovanni J (2022) Pharmacologic approaches to amino acid depletion for cancer therapy. Mol Carcinog 61(2):127–152. https://doi.org/10.1002/mc.23349. Epub 2021 Sep 17. PMID: 34534385.

Wiltschi B, Cernava T, Dennig A, Casas MG, Geier M, Gruber S, Haberbauer M, Heidinger P, Acero EH, Kratzer R, Luley C, Müller C, Pitzer J, Ribitsch D, Sauer M, Schmölzer K, Schnitzhofer W, Sensen CW, Soh J et al (2020) Enzymes revolutionize the bioproduction of value-added compounds: from enzyme discovery to special applications. Biotechnol Adv 40:107520. https://doi.org/10.1016/j.biotechadv.2020.107520

Wu J, Wang L, Sun Y, Lv S, Wu J, Zheng L et al (2025) Rational design strategy for thermostability enhancement of protein-glutaminase and investigation of the underlying mechanisms. Int J Biol Macromol 306:141580

Youssef GA, Zaid MS, Youssef AS, El-Aassar S (2025) Biochemical and biological evaluation of L-glutaminase from Aspergillus tamarii AUMC 10198 via solid-state fermentation. Microb Cell Fact 24(1). https://doi.org/10.1186/s12934-025-02802-0

Yvon M, Rijnen L (2001) Cheese flavour formation by amino acid catabolism. Int Dairy J 11(4):185. https://doi.org/10.1016/s0958-6946(01)00049-8

Zhang Z, Jin T, Xie X, Ban X, Li C, Hong Y, Cheng L, Gu Z, Li Z (2019) Structure of maltotetraose-forming amylase from Pseudomonas saccharophila STB07 provides insights into its product specificity. Int J Biol Macromol 154:1303. https://doi.org/10.1016/j.ijbiomac.2019.11.006

Zhang D, Hua Z, Li Z (2024) The role of glutamate and glutamine metabolism and related transporters in nerve cells. CNS Neurosci Ther 30(2):e14617. https://doi.org/10.1111/cns.14617. PMID: 38358002; PMCID: PMC10867874.

Zhou Y, Shen J, Chi H, Zhu X, Lu Z, Lu F, Zhu P (2024) Rational engineering and insight for a L-glutaminase activity reduced type II L-asparaginase from Bacillus licheniformis and its antileukemic activity in vitro. Int J Biol Macromol 257:128690

Zou H, Li L, Zhang T, Shi M, Zhang N, Huang J, Xian M (2018) Biosynthesis and biotechnological application of non-canonical amino acids: Complex and unclear. Biotechnol Adv 36(7):1917–1927. https://doi.org/10.1016/j.biotechadv.2018.07.008. Epub 2018 Jul 29. PMID: 30063950

9 Applications and Scope of Microbial L-Asparaginase

Rachna Shashikant Jadhav, Himanshu Sharma, and Saurabh Kulshrestha

Abstract

Microbial L-asparaginase is an amidohydrolase enzyme with significant applications in medicine and industry. It is a critical component in the treatment of acute lymphoblastic leukemia (ALL), due to its anti-leukemic properties. Beyond its therapeutic use, the enzyme is widely used in food processing to reduce the formation of acrylamide, a known carcinogen in carbohydrate-rich foods when heated at high temperatures. The enzyme is primarily sourced from microbes, which offers benefits in terms of yield and scalability. Recent advancements in enzyme engineering and novel delivery formulations, including the use of nanotechnology, aim to improve its stability and reduce its immunogenicity. The future of L-asparaginase is being shaped by innovative approaches such as CRISPR-based engineering and synthetic biology, which promise to optimize its production and utility. This chapter provides a detailed analysis of the biosynthetic potential of several microbes, recent advances in enzyme engineering, and novel delivery formulations, with the aim of elaborating on improvements to its stability and reducing immunogenicity.

Keywords

Amidohydrolase · Health · Acute lymphoblastic leukemia · Enzyme · Synthetic biology

R. S. Jadhav · S. Kulshrestha (✉)
Faculty of Applied Sciences and Biotechnology, Shoolini University, Solan, Himachal Pradesh, India
e-mail: sourabhkulshreshtha@shooliniuniversity.com

H. Sharma (✉)
Amity School of Biological Sciences, Amity University Punjab, Mohali, Punjab, India

© The Author(s), under exclusive license to Springer Nature Singapore Pte Ltd. 2026

V. Sharma, R. Salwan (eds.), *Pharmaceutically Relevant Microbial Enzymes*,
https://doi.org/10.1007/978-981-95-4941-2_9

9.1 Introduction

L-Asparaginase (ASNase) is an enzyme that has been thoroughly studied for its anticancer effects against a variety of lymphoproliferative diseases such as acute lymphoblastic leukemia (ALL) and Hodgkin's lymphoma (Egler et al. 2016). The enzyme converts L-asparagine into aspartic acid and ammonia, responsible for its anticancer properties. The enzyme is not only commonly used in the medicinal field but also in food industries to prevent the formation of acrylamide, a known carcinogen in foods rich in carbohydrates that are heated at higher temperatures (Muneer et al. 2020). A number of microbes like bacteria, fungi, algae, and even plants, produce the L-asparaginase enzyme. L-Asparaginase (ASNase) is a key therapeutic agent for cancers that depend on the amino acid asparagine, particularly acute lymphoblastic leukemia (ALL). The enzyme is found in a wide range of organisms, from bacteria and fungi to plants and animals. The current FDA-approved treatments for ALL are derived from bacterial sources, specifically *Escherichia coli* and *Erwinia chrysanthemi.* To enhance the effectiveness and reduce the side effects of ASNase, scientists are actively exploring new avenues. These include discovering and studying novel natural sources of ASNase, improving the production of recombinant versions of the enzyme, and using techniques like chemical modifications, immobilization, and nanoencapsulation. Furthermore, computational (in silico) studies are being used to gain a deeper understanding of the enzyme's function and to design more efficient and safer therapeutic options (Darvishi et al. 2022). The L-asparaginase is recognized for its various uses, including the breakdown of acrylamide in food that requires it to be heat stable, or thermostable. For this, thermostability feature, *Bacillus subtilis* was isolated from the hot springs of Tapovan, a natural habitat for heat-loving organisms, and screened for the ability to produce a thermostable ASPN enzyme (Sundaram et al. 2024).

Discovered in the early 1900s, its physiological function was unknown. However, in 1950s, a pivotal study showed that guinea pig serum contained anticancer properties, which could deplete L-asparagine and kill tumor cells. The study by Kidd and his colleagues, in 1961, demonstrated the therapeutic potential of L-asparaginase against ALL in mice sparked immediate clinical interest in the enzyme. Furthermore, the enzyme was then isolated from *Escherichia coli* (Cedar and Schwartz 1968) and *Erwinia chrysanthemi* (Nguyen et al. 2016), which laid a further foundation for advanced studies into its potential as a chemotherapeutic agent.

9.2 Classification of L-Asparaginases

L-Asparaginases are primarily classified based on the origin. The L-asparaginases from bacteria and fungi are the most common due to their high yield and ease of production, such as from *Escherichia coli* and *Erwinia chrysanthemi*. On the other side, L-asparaginases from plant and animal are commercially less viable. The L-asparaginases of bacteria are further divided into type I and II based on location and affinity for L-asparagine. The type I L-asparaginase is a low-affinity enzyme present in the cytoplasm, whereas the type II L-asparaginase have been reported from as high-affinity periplasmic enzyme. Another critical distinction is related to substrate specificity. Glutaminase hydrolyzes L-glutamine, which can lead to unwanted side effects in patients, whereas glutaminase-free enzymes are highly specific to L-asparagine, making them a safer therapeutic target for future development.

Based on its source of origin, bacterial L-asparaginases are categorized into two subtypes, namely, the L-asparaginase from *Escherichia coli* and *Erwinia chrysanthemi*. Apart from these, the genus such as *Aeromonas* spp., *Acinetobacter baumannii* (Muslim 2014), *Acinetobacter glutaminasifican* (Edman et al. 1987), *Bacillus pumilus*, *B.s pseudomycoides* (Joshi and Kulkarni 2016), *B.s licheniformis*, *B. circulans* (Gulati et al. 1997), *B. firmus* (Rudrapati and Audipudi 2017), *Enterobacter aerogenes* (Baskar et al. 2013), *E. carotovora* (Deokar et al. 2010), *Paenibacillus barengoltzii* (Qeshmi et al. 2018), *Corynebacterium glutamicum* (Ahmad et al. 2012), *P. denitriformis* (Joshi and Kulkarni 2016), *Pseudomonas aeruginosa* (Kamble et al. 2012), *Pyrococcus furiosus* (Jayam and Kannan 2014), *Pyrococcus yayanosii* (Li et al. 2018), *Staphylococcus capitis* (Paglla et al. 2013), *Streptomyces gulbargensis* (Naveena et al. 2012), *S. venezuelae* (Vidya et al. 2011), *S. ginsengisoli* (Deshpande et al. 2014), *S. alkaliphilus*, *Staphylococcus roseus* (Jha et al. 2012), *S. iranensis* (Dharmaraj 2011), *Thermococcus kodakaraensis* (Chohan and Rashid 2013), *T. gammatolerans* (Xue et al. 2014), and *Thermus thermophilus* (Prista et al. 2001a) are documented to produce L-asparaginase.

(i) *Escherichia coli*: One of the most popular microorganisms for producing recombinant microbial L-asparaginase, more especially the Type II periplasmic enzyme, is *Escherichia coli*. According to Srikhanta et al. (2013), this was the first microbe authorized for use in medicinal applications of the enzyme formulation. *Escherichia coli*-produced L-asparaginase has been investigated utilizing recombinant production techniques and has a well-defined enzyme structure and high catalytic activity (Castro et al. 2021; Maggi et al. 2021). However, despite these advantages, the limitations include glutaminase activity that is immunogenic, which can cause allergic reactions and other negative effects (Kafkewitz and Bendich 1983). PEGylation and mutant strains of *Escherichia coli* have been developed to increase the stability of L-asparaginase and decrease the toxicity of the enzyme. ONCASPAR and pegaspargase are the two forms of *Escherichia coli* L-asparaginase (Meneguetti et al. 2019a, b). *Escherichia coli* has also been used as a vector for overexpression and large-scale production of the L-asparaginase enzyme by Kishore et al. (2015). Batch culture and fed-batch culture are the most common bioreactor methods for the production of *Escherichia coli*-based L-asparaginase enzyme (Barros et al. 2021).

(ii) *Erwinia chrysanthemi*: After *Escherichia coli*, *Erwinia chrysanthemi* is another bacterium that was approved for the production of the L-asparaginase enzyme. It is immunologically distinct from the enzyme obtained from *Escherichia coli* (Salzer et al. 2014). Erwinase is the native L-asparaginase enzyme obtained from *Erwinia chrysanthemi* and a second-line approved treatment for children and adolescents with ALL, but they face hypersensitivity issues or neutralizing antibody development due to *Escherichia coli*-based L-asparaginase. But considering the short half-life in vivo of the enzyme, patients frequently need to be administered the same. Thus, Modi and Gervais (2022) developed a native *Erwinia chrysanthemi*-based L-asparaginase by attaching it covalently to polyethylene glycol (PEG), so that its half-life in vivo is extended. Spectrila is another commercially marketed L-asparaginase that is known to have a lower chance of allergies and a better half-life in vivo (Schnuchel et al. 2023).

(iii) *Streptomyces* spp.: Actinomycetes belonging to *Streptomyces* spp. like *Streptomyces albidoflavus*, *Streptomyces griseus*, *Streptomyces lacticiproducens*, and *Streptomyces* spp. HB2AG. *Streptomyces* spp. produce extracellular L-asparaginase, which simplifies the purification process. They produce the enzyme with lower glutaminase activity and better pH and thermal stability. L-asparaginase enzymes produced from *Streptomyces* spp. are suitable for use in the therapeutic and food industries. Submerged fermentation is used for *Streptomyces albidoflavus*-based L-asparaginase as studied by Narayana et al. (2008). Another novel study on *Streptomyces* spp. HB2AG was studied, wherein Rath et al. (2023) isolated it from a sediment near Chilika Lake in Odisha, India, for which whole genome analysis revealed the presence of asparaginase synthase in the genome. DeJong (1972) conducted studies on the production of isolation of L-asparaginase from *Streptomyces griseus*, which was found to be comparable and equally good as that obtained from *Escherichia coli.* Another study by Arévalo-Tristancho et al. (2019) isolated *Streptomyces lacticiproducens* from the riverbank of Arauca, Colombia, which was found to produce L-asparaginase.

(iv) *Bacillus* spp.: *Bacillus* spp., like *Bacillus subtilis*, *Bacillus licheniformis*, *Bacillus circulans*, *Bacillus niacini*, and *Bacillus velezensis*, have been commonly explored for L-asparaginase production. Several studies have proven that the enzymes, when extracted from these species, are non-pathogenic, have a good secretion system, and produce thermostable variants. *Bacillus*-derived L-asparaginases can be used for recombinant production and as co-expression systems (Safary et al. 2019). Moreover, they are environment-friendly and used widely for food-processing applications, apart from therapeutic applications. Alrumman et al. (2019) isolated 40 strains from the Red Sea, Saudi Arabia, and found that plenty of those isolates were able to produce glutaminase-free L-asparaginase. Another study by Wahab et al. (2024) focused on the isolation of *Bacillus niacini* from a marine sediment that is able to produce the extracellular L-asparaginase enzyme. Mahajan et al. (2012) optimized conditions for the production of L-asparaginase using *Bacillus licheniformis* with low glutaminase activity. Abdelrazek et al. (2019) conducted experimental and bioinformatics studies to scan several soil samples and concluded that *Bacillus licheniformis* gave a high yield of L-asparaginase enzyme, whereas a study by Mostafa et al. (2019) on marine sediments enhanced the production of glutaminase-free L-asparaginase enzyme using *Bacillus velezensis.*

(v) *Pseudomonas* spp.: In the *Pseudomonas* spp., *Pseudomonas aeruginosa and Pseudomonas fluorescens* are known to produce L-asparaginase with unique properties compared to other bacterial L-asparaginases. These genus isolates can produce extracellular as well as intracellular isoforms of L-asparaginase (Badoei-Dalfard 2015; Jana et al. 2024). However, despite their unique abilities, *Pseudomonas aeruginosa* strains are considered opportunistic pathogens, which limits their therapeutic usage unless they are expressed in non-pathogenic hosts. Owing to this limitation, research being conducted lately using these species focuses on recombinant expression in safer vectors for enzyme production at an industrial level. Kishore et al. (2015) cloned *Pseudomonas fluorescens*-based L-asparaginase and overexpressed it in *Escherichia coli* to mitigate the pathogenic potential of *Pseudomonas.* Another study by Fatima et al. (2019) isolated *Pseudomonas aeruginosa* from garden soil samples and found out its ability to produce the L-asparaginase enzyme.

(vi) *Aspergillus* spp.: *Aspergillus* spp. like *Aspergillus niger* and *Aspergillus terreus* are other examples of microbes that are able to produce the L-asparaginase enzyme. They have GRAS status (Generally recognized as safe) and are known to produce extracellular enzymes. Most commonly, they are employed in the food industry for the reduction of acrylamide content in fried or baked food products (da Cunha et al. 2019). They have slightly lower enzyme activity compared to the enzyme obtained from the bacterial species mentioned above, and recent research has focused on downstream processing but is complex owing to fungal biomass. Rabelo et al. (2025) have employed the solid-state fermentation method to produce L-asparaginase with *Pereskia aculeata*, using *Aspergillus niger.* Kassab (2023) produced the L-asparaginase enzyme employing *Aspergillus niger* from soil environments in Egypt, which showed no hypersensitivity reactions, glutaminase-like effects, and exhibited better bioactivity compared to bacterial L-asparaginases. Another study by Hassan et al. (2018) extracted L-asparaginase from marine *Aspergillus terreus* and found that it exhibited anticancer activity against HCT-116, Hep-G2, and MCF-7 cell lines in vitro.

Fungal L-Asparaginase L-asparaginases from fungi have been derived from *Aspergillus, Fusarium, and Penicillium* spp., owing to their varied isoforms and potential for applications in the food industry and low toxicity. Fungi such as *A. alliaceus* (Gulati et al. 1997), *A. tamari* (Sanjotha and Manawadi 2017), *A. terreus* (Sarquis et al. 2004), *A. tubingensis* (Gonçalves et al. 2017), *A. niger*, *A. nidulans* (Baskar and Renganathan 2011), and *A. fumigatus* (Benchamin et al. 2019). Parashiva et al. (2023) have been explored for L-asparaginase enzyme.

L-Asparaginase from Endophytes Besides free living bacterial and fungal population, another subcategory of microorganism representing endophytes represent plant-resident flora that live inside plant tissues and are capable of producing a variety of bioactive substances (Nelson and Nelson 2012). Chow and Ting (2015) conducted a study on *Pereskia bleo* (seven-star needle), *Oldenlandia diffusa* (Chinese herb), *Murraya koenigii* (curry plant), and *Cymbopogon citratus* (lemon grass), four common medicinal plants in China, from which endophytic fungi were isolated for the production of L-asparaginase. L-asparaginase is also found in remarkably high concentrations in *Tamarindus indica* and *Vigna radiata* (Fabaceae) (Aisha et al. 2020). The L-asparaginase enzyme, which is found in crops like soybeans and legumes, has also been investigated for extraction from *Solanum nigrum* (Shanmugaprakash et al. 2015). Although the yield of the enzyme from plants is comparatively less, and purification is a challenge, they are used to extract L-asparaginase.

9.3 Mechanism of L-Asparaginase

Asparaginases, particularly the Type II L-asparaginases from *E. coli* and *Dickeya chrysanthemi*, utilize a unique double-displacement (ping-pong) mechanism to catalyze the hydrolysis of L-asparagine into L-aspartate and ammonia (Saeed et al. 2021). This process involves a series of steps where the enzyme's active site acts as both a nucleophile and a leaving group. The catalytic reaction is initiated by a weak nucleophile, a threonine residue in the enzyme's active site. This threonine attacks the carbonyl carbon of the L-asparagine substrate, forming a temporary covalent intermediate. This intermediate is stabilized by a previously undescribed motif within the enzyme's structure. Following this, the ammonia molecule is released. A water molecule then enters the active site, where it is activated to attack the covalent intermediate. This leads to the formation of L-aspartate and regenerates the active site of the enzyme, allowing it to begin the cycle anew. The enzyme's ability to perform this reaction without the direct assistance of a general base, a feature common in many other hydrolytic enzymes, makes its mechanism particularly notable. For over four decades, two bacterial Type II L-asparaginases, derived from *Escherichia coli* and *Dickeya chrysanthemi*, have been crucial in treating childhood leukemias and lymphomas. Despite their widespread use, the exact mechanism of their catalytic reaction has remained a subject of debate. This report presents a detailed model for the reaction catalyzed by the *E. coli* enzyme (EcAII), combining new structural and biochemical findings with prior research. The study provides strong evidence that EcAII uses a double-displacement (ping-pong) mechanism involving a covalent intermediate. This process has several unique features: a weak nucleophile, threonine, initiates the reaction without a direct general base, though a more distant one is identified. Additionally, the tetrahedral intermediates that form during catalysis are stabilized by a previously unknown motif. While this catalytic scheme was developed using only EcAII, the researchers suggest that this novel mechanism for enzymatic hydrolysis could apply to most, if not all, L-asparaginases (Lubkowski et al. 2020).

9.4 Structural Organization of Asparaginases

Asparaginases are a diverse family of enzymes, but many share a similar foundational structure, particularly the medically important bacterial Type II L-asparaginases. These enzymes, which include those from *Escherichia coli* (EcAII) and *Dickeya chrysanthemi*, are typically large, complex proteins (Koval et al. 2021). The most common structural form for these bacterial enzymes is a homotetramer, meaning it is composed of four identical subunits (Aghaiypour et al. 2001). Each of these subunits, or monomers, is about 330 amino acids long and is folded into two distinct domains: an N-terminal domain and a C-terminal domain, connected by a linker region (Swain et al. 1993a, b). The active sites of the enzyme are located at the interface between these domains and subunits, where they form a catalytic site that binds and hydrolyzes the L-asparagine substrate. The overall quaternary structure is described as a "dimer of two intimate dimers" (Cerofolini et al. 2019), which creates a D2-symmetric arrangement. While Type II asparaginases are well established as homotetramers, other asparaginase types have different structural arrangements. For example, some Type I asparaginases may exist as dimers or even function as hexamers, which are trimers of dimers, in some species (Pritsa and Kyriakidis 2001a, b). Furthermore, the structural classification of asparaginases is divided into bacterial-type, plant-type, and Rhizobium etli-type, highlighting the evolutionary and structural diversity of these enzymes across different life forms (Michalska et al. 2006).

9.5 Biochemical Characterization of L-Asparaginase

L-asparaginase catalyzes the hydrolysis of L-asparagine into L-aspartic acid and ammonia. The kinetic parameters that are essential to be considered are the Michaelis constant, maximum enzyme velocity, and its catalytic efficiency. The Michaelis constant is denoted by K_m, maximum enzyme velocity by V_{max}, and catalytic efficiency by K_{cat}/K_m. The enzyme's affinity for L-asparagine substrate is indicated by K_m. The K_m of therapeutic enzymes is low. The V_{max} varies with the source and isoform, whereas the catalytic efficiency is frequently maximized by changes in genetic sequence or the selection of strains with superior performance. The best combination to reduce the side effects of the enzyme in therapy is high specificity of L-asparagine and low affinity and low activity toward L-glutamine. Aktar et al. (2023) discuss the structural and kinetic properties of pharmaceuticals based on L-asparaginase.

Thermostable microbial L-asparaginases are derived from strains of *Streptomyces* or *Bacillus*, which exhibit greater stability at temperatures between 40 and 60 °C, whereas hyperthermophilic strains work efficiently at 85–100 °C (Jia et al. 2021). This makes them popular for food and bioprocessing applications. Sundaram et al. (2024) have isolated *Bacillus subtilis* from the hot springs of Tapovan to screen a thermostable L-asparaginase, thereby recognizing its varied applications in the food industry for acrylamide degradation. L-asparaginases that are alkaline-tolerant are favored in several industries. Optimal activity of the enzyme, however, is at pH 6.5–8.5. Microbial L-asparaginases are also known to be stable at pH 5–9 (Honfoga et al. 2025). Certain L-asparaginases are also halophiles, which thrive well in high salt content environments and exhibit good efficiency. They are employed in applications related to saline wastewater and marine environments. Ebrahiminezhad et al. (2011) have isolated halophilic *Bacillus* spp. BCCS 034 from the Maharloo salt lake, Iran, which produced a high yield of extracellular L-asparaginase, with novel immunonological properties.

Employing the PEGylation method, which involves a covalent bond of polyethylene glycol, helps mask the immunogenic areas and extends the circulatory half-life of the enzyme. Meneguetti et al. (2019a, b) developed a site-specific N-terminus PEGylation protocol for L-asparaginase production, which led to the formation of monoPEG-L-asparaginase. This version of the PEGylated enzyme exhibited more enzymatic stability for more days than non-PEGylated L-asparaginase. It showed resistance to plasma proteases like asparaginyl endopeptidases and cathepsin B. Furthermore, they were also found to possess anti-leukemic activity in the cell lines MOLT-4 and REH in vitro. Chimeric fusion proteins, too, help in increasing the enzyme solubility and secretion, as discussed earlier. Apart from this, ML models and computational designs make structural variant prediction with improved stability and therapeutic applications easier, as studied by Pedroso et al. (2025). Owing to recent advances in nanotechnology and nanoformulations, the encapsulation of enzymes in nanoparticles improves delivery and reduces their immunological recognition. Researchers have employed silk sericin microparticles (Zhang et al. 2004), biopolymers like chitosan (Bahreini et al. 2014), supramolecular structures like polymerosomes (Bueno et al. 2020), polyion complex vesicles (Kurinomaru and Shiraki 2015), virus-like particles (Hill et al. 2018), and red blood cells (Domenech et al. 2011; Rossi et al. 2020) for encapsulation of L-asparaginases.

9.6 Application of Mirobial L-Asparaginase

Although fungi and plants can be used to extract L-asparaginase, microbes remain the preferred source for the production of L-asparaginase at an industrial level, owing to their high yield and cost-effective benefits. Bacterial strains like *Escherichia coli*, *Bacillus subtilis*, and several actinomycetes species produce significant amounts of L-asparaginase through different types of fermentation methods (Muneer et al. 2020). Microbes can be used more conveniently due to the possibility of fermentation in bioreactors for large-scale enzyme extraction (Liu and Zajic 1973). Another benefit of using microbes for the production of enzymes, especially L-asparaginase, is that bacteria are easy to modify at a genetic level, which can help increase enzyme activity and yield (Meena et al. 2015). Additionally, compared to other sources like plants and fungi for the production of the L-asparaginase enzyme, microbial production is more environmentally friendly, as it is mostly based on green synthesis techniques. These features make microbial L-asparaginase an essential part of enzyme biotechnology for both industrial and medicinal applications.

In terms of cancer therapy, the mechanism of action of the L-asparaginase enzyme exploits the metabolic requirement of cancer cells. Cancer cells like lymphoblasts cannot produce L-asparagine on their own due to low levels of asparagine synthetase and have a need to obtain it from other external sources. It is the L-asparaginase enzyme that exploits their metabolic vulnerability to inhibit protein synthesis, apoptosis, and cell death by reducing extracellular asparagine. L-asparaginase is an amidohydrolase, which means it can hydrolyze not only asparagine but also glutamine (Chan et al. 2014). Plasma L-asparagine levels are rapidly reduced upon the administration of L-asparaginase, leading to inadequate intake of amino acids, inhibition of nucleotide and protein synthesis, stress to the endoplasmic reticulum, and the induction of autophagy and apoptosis (Lubkowski et al. 2020).

L-asparaginase is mostly used for tumor-specific treatments, as normal cells can synthesize asparagine de novo, so this selective deprivation has little effect on them. Furthermore, some downstream consequences of L-asparagine depletion work by disrupting the progression of the cell cycle (Babu et al. 2011), preventing the growth of new blood vessels (Yu et al. 2012), and inhibiting the mTOR signaling pathway that is important for the survival of cancer cells (Samudio and Konopleva 2013). Certain bacteria produce L-asparaginases that hydrolyze L-glutamine, which is involved in the synthesis of nucleotides and cellular metabolism. While this increases cytotoxicity, it also exacerbates other undesirable side effects like pancreatitis, neurotoxicity, and hepatotoxicity (Chan et al. 2014). However, L-asparaginases are being developed nowadays in a way that they have minimal or no glutaminase activity, so the therapeutic safety of the enzyme increases (Yu et al. 2025; Shafqat et al. 2023). On the other hand, asparaginase is also widely used in the food industry. The food industry continues to use deamination of free L-asparagine treatment in starchy food matrices like wheat and potatoes before thermal processing. Due to this, the Maillard reaction between the reducing sugars and asparagine during the baking or frying process produces less acrylamide, which is a known carcinogen (Xu et al. 2016). Several studies have shown that microbial L-asparaginases are usually tetrameric proteins, with each monomer contributing to the active site. The immunogenicity, pH tolerance, and substrate selectivity of the enzyme can be changed by modifications in the flexible loop sections, as catalysis and substrate binding are related to these regions (Swain et al. 1993a, b). Additionally, recently, the advances in site-directed mutagenesis and molecular modeling have made it feasible to design reasonable enzymes that improve the efficacy of catalysis, the consistency of the enzyme, and the preference for the synthesis of L-asparagine over L-glutamine (Pokrywka et al. 2024).

9.7 Optimization of L-Asparaginase Production

Optimizing L-asparaginase production is a critical area of research, with recent strategies focusing on a combination of genetic and bioprocess engineering to enhance yield and therapeutic properties. A primary approach involves recombinant technology, where the genes for asparaginase are expressed in host microorganisms like *E. coli*, *Bacillus subtilis*, and *Pichia pastoris* to achieve high-volume production (Shishparenok et al. 2023). This method allows for the strategic modification of the enzyme to improve its stability, reduce immunogenicity, and decrease its affinity for glutamine, which helps to mitigate side effects. Beyond genetic manipulation, fermentation process optimization remains a vital strategy. Researchers use statistical methods like response surface methodology (RSM) and Plackett–Burman design to systematically evaluate and optimize various production parameters, including substrate type, carbon and nitrogen sources, pH, temperature, and incubation time (Abhini et al. 2022). For example, studies have shown that optimizing nutrient and oxygen requirements in fed-batch cultures can significantly increase enzyme activity and productivity (Lima et al. 2020). Solid-state fermentation using low-cost substrates like wheat bran or bajra seed flour is also gaining traction as a cost-effective and efficient method for large-scale production (Abdel-Hamid et al. 2022). Collectively, these approaches are crucial for developing more efficient and safer L-asparaginase products for both clinical and food industry applications.

9.7.1 Submerged Versus Solid-State Fermentation Strategies

L-Asparaginase is produced on a large scale at the industrial level using microbial fermentation methods. The two most popular methods for the fermentation of L-asparaginase enzyme are submerged fermentation (SmF) and solid-state fermentation (SSF) (Doriya et al. 2016). Every method has pros and cons, depending on the microbial strain, the substrate, the desired product shape, and whether the application is industrial or therapeutic.

9.7.1.1 Submersion Fermentation

Submerged fermentation (SmF) is the process of cultivating microorganisms in liquid nutritional media under precisely controlled environmental conditions, including the optimal temperature, pH, agitation, and aeration. This fermentation process is the most popular way to produce the L-asparaginase enzyme for use in medicine and therapy. This method can be used to produce extracellular as well as intracellular L-asparaginase using batch, fed-batch, or continuous modes. The advantages include precise control over the fermentation parameters, ease in scaling up, automating the process thereby simplifying it, and enabling the monitoring of enzyme yield, pH, and biomass in real time. However, it has disadvantages like overconsumption of water and energy sources for agitation and aeration and chilling purposes; high media requirements; increased operating costs; and sometimes excess foam formation can lead to contamination issues (Kumar and Sobba 2013). The most widely cultured microbes using SmF are *Escherichia coli*, *Erwinia chrysanthemi*, *Pseudomonas aeruginosa*, and *Bacillus subtilis.* El-Naggar et al. (2019), have optimized fermentation conductions and produced the L-asparaginase enzymes using *Streptomyces brollosae* NEAE-115 strain by submerged fermentation method. Baskar and Ranganathan (2009), have varied eight parameters for the production of L-asparaginase using *Aspergillus terreus* MTCC 1782. Another study by Erva et al.

(2016), employed the use of a novel *Enterobacter* spp., *Enterobacter aerogenes KCTC2190/MTCC111*, for optimization of extracellular L-asparaginase production.

9.7.1.2 Solid-State Fermentation (SSF)

SSF is using solid substrate materials like agro-waste, rice husks, walnut shells, or wheat bran with little or no free water to culture microbes for fermentation. This method is widely employed for generating extracellular L-asparaginase using actinomycete strains and fungi. SSF helps stimulate the natural habitat of a number of filamentous fungi and actinomycetes, encourages sustainability, and reduces agro-industrial waste. Its advantages include increased volumetric output of extracellular L-asparaginase enzyme, minimal wastewater production, enhanced stability, enzyme activity, and low operating costs (El-Naggar et al. 2019). It is the best-suited method for the generation of GRAS enzymes from *Aspergillus niger*. Its limitations include monitoring difficulties caused by control factors such as pH, temperature, and oxygen transport, as well as issues with scale-up because of flaws in the solid matrix. It is limited to species that have low tolerance for water activity. Using *Serratia marcescens* and the conventional one factor at a time (OFAT) and response surface methodology (RSM), Ghosh et al. (2013) generated the L-asparaginase enzyme using coconut oil cake as a solid substrate for SSF. This resulted in maximum enzymatic activity and a high enzyme yield. In order to create an affordable L-asparaginase enzyme, Sharma and Mishra (2021) employed a variety of agricultural wastes as substrates for SSF utilizing *Aspergillus niger* and an artificial neural network technique. Linseed de-oiled cake, rice husk, wheat bran, maize bran, pea pod husk, and a new niger de-oiled cake were among these wastes. The main determinants thus deciding whether SmF or SSF will be beneficial for the production of L-asparaginase enzyme are final application, budget, substrate, and microbial sources. SSF is gaining popularity for both industrial and starchy-food grade enzyme applications, especially because bioprocessing becomes more economical and sustainable, whereas SmF remains the industry-standard method for fermentation to produce pharmaceutical-grade enzymes due to its superior sterility and control. However, some studies integrate both SmF and SSF methods as described by Martău et al. (2021) and Florencio et al. (2016).

9.7.2 Optimization of Culture Conditions for Maximum Yield

In order to increase the yield of microbial L-asparaginase, optimization of growth conditions, reducing production costs, and enhancing enzyme activity and stability are essential. Several physical, chemical, and nutritional factors play a vital role in influencing the production of the enzymes. Physical parameters include temperature, inoculation time, aeration, and agitation. Chemical parameters like pH, inoculum size, and inducer presence, whereas nutritional parameters like carbon and nitrogen sources, play a pivotal role in enzyme production. Additionally, the OVAT and RSM methods provide basic insights and can be widely used for optimizing multiple variables simultaneously with increased efficacy (Table 9.1).

Table 9.1 Critical fermentation parameters influencing the yield of microbial L-asparaginases

S. no.	Optimal conditions	Major characteristics	References
1	6.5–8.5	Neutral to slightly alkaline pH favors enzyme stability and activity	Narayana et al. (2008)
2	30–37 °C (*E. coli*, *Bacillus*) 25–30 °C (*Aspergillus*)	Microbial strain dependent, whether thermophilic, mesophilic, or hyperthermophilic	Distasio et al. (1982); Jia et al. (2021); Kumar and Savitha (2015)
3	Glucose, sucrose, glycerol	Energy sources, excessive amounts may cause catabolite repression	Elshafei and El-Ghonemy (2015); Sahu et al. (2007); Narayana et al. (2008)
4	Yeast extract, peptone, ammonium salts, L-asparagine	L-asparagine acts as an inducer to enzyme production	Khamna et al. (2009); Narayana et al. (2008); Sudhir et al. (2012); Amena et al. (2010)
5	0.5–1.5% (w/v)	Acts as an inducer and a nitrogen source	El-Naggar et al. (2019)
6	100–200 rpm (shake flask); 0.5–1.0 vvm (bioreactor)	Ensures overall oxygen transfer. Excessive agitation may damage cells	Jain et al. (2012)
7	24–72 h	Peak enzyme yield between log and early stationary phase	Isaac and Abu-Tahon (2016); Pallem (2019)
8	5–10% (v/v)	Less or excess yield may affect enzyme yield and biomass	Singh and Srivastava (2013)
9	SmF or SSF	Choice dependent on strain type and application of enzyme	Doriya et al. (2016)
10	RSM, Taguchi, DoE	Used for multivariate optimization and interaction analysis	Abhini et al. (2022); Ashok et al. (2019); Jha et al. (2014)

The growth conditions must be tailored to the microbial strain and production platform (SmF or SSF) in order to maximize the yield of L-asparaginase. Systematic optimization using statistical models not only boosts productivity but also ensures economic viability and reproducibility for industrial manufacturing.

9.8 Genetic and Metabolic Engineering Approaches

Modern biotechnology research has significantly enhanced the production, activity, and therapeutic potential of L-asparaginase. Genetic engineering and metabolic pathway optimization is altogether crucial for improving enzyme productivity and specificity, as well as reducing glutaminase activity. By altering an organism's genetic material, recombinant DNA technology can increase the production of enzymes and change their characteristics, increasing their usefulness for a variety of applications. L-asparaginase is produced using recombinant DNA technology. Shakambari et al. (2018) employed this technique to overexpress the protein by cloning the genetic material that causes *Bacillus tequilensis* PV9W to produce the L-asparaginase enzyme into a different heterologous bacterial host. The enzyme produced exhibited better stability, trypsin tolerance, and higher thermostability. The same method was applied in a different study by Ferrara et al. (2006) utilizing a recombinant strain of *Pichia pastoris* that carried the *Saccharomyces cerevisiae* ASP3 gene, which led to a greater production of the L-asparaginase enzyme. Researchers set out to develop a recombinant L-Asparaginase (ASNase) in human cells to address the side effects, such as hypersensitivity and inactivation, associated with the bacterial-origin enzymes currently used to treat acute lymphoblastic leukemia (ALL). Their objective was to investigate whether producing the enzyme in human cells would lead to post-translational modifications and, if so, what the consequences would be. The study successfully expressed a recombinant ASNase in human cells, finding it had a molecular weight of 60 kDa, which is significantly larger than the 35 kDa enzyme produced in *Escherichia coli*. Further analysis confirmed that this increased size was due to N-glycosylation, where the human cells added sugar molecules (glycans) to the protein. This glycosylated ASNase remained active under physiological conditions (pH and temperature). The findings are promising because glycosylation can potentially reduce the enzyme's antigenicity by masking its epitopes, which are the parts of the protein that trigger an immune response. This research could pave the way for a new ASNase treatment option for patients with ALL who experience severe side effects from existing bacterial-derived enzymes (Dantas et al. 2019).

Limited supply of *Erwinia chrysanthemi*-derived asparaginase (ERW), a key treatment for patients with hypersensitivity to other forms of the enzyme, has created a need for alternatives. Recombinant technology offers a solution by programming cells to produce a new asparaginase. One such product is JZP-458, a recombinant *Erwinia* asparaginase developed using a *Pseudomonas fluorescens* expression platform. JZP-458 shares the same amino acid sequence as ERW, has comparable activity in laboratory tests, and, importantly, does not cause an immune reaction in patients sensitive to *E. coli*-derived asparaginases. The efficient production of JZP-458 would provide a reliable new supply of this vital medication, ensuring patients with hypersensitivity have access to the treatment they need (Maese et al. 2021).

A molecular biology-based method called site-directed mutagenesis enables precise, targeted modifications to the amino acid sequence of the enzyme, potentially improving its functionality. Certain amino acid residues in the active site change to increase the specificity of the substrate, enhance the pH tolerance and thermostability of the enzyme, and reduce glutaminase activity. A recent study by Pokrywka et al. (2025) examined the active site of an L-asparaginase using mutagenesis of Ser-Lys tandems in *Rhizobium etli*. CRISPR-Cas systems edit a genome. This system can be applied in bacteria as well, which can either assist in the removal of inhibitory regulatory elements or increase native gene expression. It enables the development of microbial strains that have low glutaminase activity and high yields of enzymes such as L-asparaginases. *Erwinia persicina* BST187's CRISPR-Cas9 system was used in a recent study by Cheng et al. (2025) to create a quick and effective genome-editing technique. They concentrated on enhancing the structure of sgRNA and increasing its levels of expression in order to attain a high level of gene editing precision. This study provided a framework to create effective gene-editing tools for microbial strains. This kind of study demonstrates its potential as a promising platform for the production of natural products. When L-asparaginase is fused with signal peptides, it promotes extracellular secretion and reduces the need for cell lysis. It also streamlines the downstream processing. Oza et al. (2011) used a His-Tag fusion protein purification kit to purify recombinant L-asparaginase from *Withania somnifera.* The study reported a high enzymatic activity of L-asparaginase.

In this way, genetic studies and metabolic engineering enable the practical and logical development of next-gen L-asparaginases with improved performance, appropriate for specific uses in industries, food, and therapeutic applications. These methods not only boost efficiency but also open up new avenues for enzyme modification and synthetic biology applications.

The discovery, innovations, and optimization of microbial L-asparaginases are being revolutionized by the recent developments in synthetic biology, and owing to advances in artificial intelligence (AI), microbial L-asparaginases can be further optimized. Machine learning (ML) models enable analysis of large datasets to predict enzyme variations with improved activity, reduced glutaminase activity, or reduced immunogenicity, and enhanced enzyme stability. Deep learning enables rational mutagenesis by accelerating the prediction of protein structures. El-Naggar et al. (2024) have conducted optimization of extracellular L-asparaginase from *Streptomyces violaceoruber* using SSF, which was free of L-glutaminase, by employing AI.

Consequently, AI has made it possible to create microbial L-asparaginases in a logical and predictive manner as opposed to depending on trial-and-error. This could result in more efficient and quicker-to-develop tailored enzymes for use in food, medicine, and other industrial applications.

The therapeutic efficacy and industrial adaptability of microbial L-asparaginases can be improved further if their biochemical and structural properties are well studied. Enzymes differ in structure, stability, isoforms, and catalytic behaviors, all of which affect their efficacy. Microbial L-asparaginases are primarily classified into two types. Type I enzymes are cytosolic enzymes; that is, they are found in the cytoplasm. They exhibit low affinity toward L-asparagine and have a relatively high specific activity toward L-glutamine. Most Type I microbial L-asparaginases derived from thermophiles are homodimers with around 70 kDa molecular mass. Type II enzymes, on the other hand, are periplasmic enzymes, found in the periplasmic space. They exhibit low to negligible specific activity against L-glutamine, in addition to a high specificity against L-asparagine. These are more valuable for medicinal applications. Some organisms are also known to produce Type III, commonly known as plant L-asparaginases, which completely differ in structure and sequence from Type I and Type II and are less studied for clinical use (Qeshmi et al. 2018; Pourhossein and Korbekandi 2014; Shi et al. 2017).

9.9 Conclusion

The microbial L-asparaginase enzyme is a critical asset in both the medical and food industries, offering significant benefits over enzymes from other sources like plants and animals due to its high yield and cost-effectiveness. In medical applications, L-asparaginase is a cornerstone of acute lymphoblastic leukemia (ALL) treatment, leveraging the metabolic vulnerability of cancer cells to inhibit protein synthesis and induce cell death. This chapter highlights the development of new formulations, such as PEGylated L-asparaginase, and innovative production methods that minimize glutaminase activity, thereby reducing unwanted side effects like pancreatitis and neurotoxicity while extending the enzyme's half-life. In the food industry, L-asparaginase is essential for reducing acrylamide content in starchy foods, a key application for ensuring food safety. The production of the enzyme has been optimized through genetic and bioprocess engineering, including the use of recombinant DNA technology and statistical methods like response surface methodology (RSM). This chapter also compares submerged fermentation (SmF) and solid-state fermentation (SSF), noting that while SmF is the industry standard for pharmaceutical-grade enzymes, SSF is becoming more popular for industrial applications due to its economic and sustainable benefits. Continued research and advancements in genetic engineering and nanotechnology promise to further enhance the therapeutic efficacy, stability, and production of L-asparaginase, solidifying its role as an indispensable enzyme in biotechnology.

References

Abdel-Hamid NS, Abdel-Khalek HH, Ramadan E, Mattar ZA, Abou-Taleb KA (2022) Optimization of L-asparaginase production from Fusarium oxysporum F-S3 using irradiated pomegranate peel under solid-state fermentation. Egypt J Chem 65(6):381–397. https://doi.org/10.21608/ejchem.2021.104129.4817

Abdelrazek NA, Elkhatib WF, Raafat MM et al (2019) Experimental and bioinformatics study for production of L-asparaginase from *Bacillus licheniformis*: a promising enzyme for medical application. AMB Express 9:39. https://doi.org/10.1186/s13568-019-0751-3

Abhini KN, Rajan AB, Fathimathu Zuhara K, Sebastian D (2022) Response surface methodological optimization of L-asparaginase production from the medicinal plant endophyte *Acinetobacter baumannii* ZAS1. J Genet Eng Biotechnol 20(1):22. https://doi.org/10.1186/s43141-022-00309-4

Aghaiypour A, Ghassempour A, Mofidian MA, Siahpoosh M (2001) The crystal structure of L-asparaginase from *Erwinia chrysanthemi*. Acta Crystallogr D Biol Crystallogr 57(12):1835–1842

Ahmad N, Pandit N, Maheshwari S (2012) L-asparaginase gene—a therapeutic approach towards drugs for cancer cell. Int J Biosci 2(4):1–11

Aisha A, Zia MA, Asger M, Muhammad F (2020) L-asparaginase, acrylamide quenching enzyme production from leaves of Tamarindus indica and seeds of Vigna radiata–Fabaceae. Pak J Bot 1(1):243–249

Aktar BY, Georgakis N, Labrou N, Turunen O, Binay B (2023) Comparative structural and kinetic study for development of a novel candidate L-asparaginase based pharmaceutical. Biochem Eng J 191:108806, ISSN 1369-703X. https://doi.org/10.1016/j.bej.2023.108806

Alrumman SA, Mostafa YS, Al-Izran KA et al (2019) Production and anticancer activity of an L-asparaginase from *Bacillus licheniformis* isolated from the Red Sea, Saudi Arabia. Sci Rep 9:3756. https://doi.org/10.1038/s41598-019-40512-x

Amena S, Vishalakshi N, Prabhakar M, Dayanand A, Lingappa K (2010) Production, purification and characterization of L-asparaginase from *Streptomyces gulbargensis*. Braz J Microbiol 41:173–178

Arévalo-Tristancho E, Díaz L, Cortázar J, Valero M (2019) Production and characterization of L-asparaginases of *Streptomyces* isolated from the Arauca riverbank (Colombia). Open Microbiol J 13:204–215. https://doi.org/10.2174/1874285801913010204

Ashok A, Doriya K, Rao JV, Qureshi A, Tiwari AK, Kumar DS (2019) Microbes producing L-asparaginase free of glutaminase and urease isolated from extreme locations of Antarctic soil and moss. Sci Rep 9(1):1423. https://doi.org/10.1038/s41598-018-38094-1

Babu PJ, Rajesh KV, Prasad PD, Manjusha K, Rani KV (2011) Role of L-asparaginase as an effective chemotherapeutic agent for acute lymphoblastic leukemia in children's. Biosci Biotechnol Res Asia 8(2):641–651

Badoei-Dalfard A (2015) Purification and characterization of L-asparaginase from Pseudomonas aeruginosa strain SN004: production optimization by statistical methods. Biocatal Agric Biotechnol 4(3):388–397, ISSN 1878-8181. https://doi.org/10.1016/j.bcab.2015.06.007

Bahreini E, Aghaiypour K, Abbasalipourkabir R, Goodarzi MT, Saidijam M (2014) Preparation and nanoencapsulation of L-asparaginase II in chitosan-tripolyphosphate nanoparticles and in vitro release study. Nanoscale Res Lett 9:340

Barros T, Brumano L, Freitas M, Pessoa A Jr, Parachin N, Magalhães PO (2021) Development of processes for recombinant L-asparaginase II production by *Escherichia coli* Bl21 (De3): from shaker to bioreactors. Pharmaceutics 13(1):14. https://doi.org/10.3390/pharmaceutics13010014

Baskar G, Ranganathan S (2009) Statistical screening of process variables for the production of L-asparaginase from cornflour by Aspergillus terreus MTCC 1782 in submerged fermentation. Indian J Sci Technol 2(5):45–48

Baskar G, Renganathan S (2011) Optimization of media components and operating conditions for exogenous production of fungal L-asparaginase. Chiang Mai J Sci 38(2):270–279

Baskar G, Rajasekar V, Renganathan S (2013) Modeling and optimization of L-asparaginase production by *Enterobacter aerogenes* using artificial neural network linked genetic algorithm. Int J Chem Eng Appl 2:98–100. https://doi.org/10.7763/ijcea.2011.v2.83

Benchamin D, Sreejai R, Sujitha S, Albert C (2019) Anti-proliferative activity of L-asparaginase enzyme from fungi on breast cancer. J Pharmacogn Phytochem 8(1):407–410

Bueno CZ, Apolinário AC, Duro-Castano A, Poma A, Pessoa A Jr, Rangel-Yagui CO, Battaglia G (2020) L-asparaginase encapsulation into asymmetric permeable polymersomes. ACS Macro Lett 9:1471–1477

Castro D, Marques ASC, Almeida MR, de Paiva GB, Bento HBS, Pedrolli DB, Freire MG, Tavares APM, Santos-Ebinuma VC (2021) L-asparaginase production review: bioprocess design and biochemical characteristics. Appl Microbiol Biotechnol 105(11):4515–4534. https://doi.org/10.1007/s00253-021-11359-y

Cedar H, Schwartz JH (1968) Production of L-asparaginase II by *Escherichia coli*. J Bacteriol 96(6):2043–2048. https://doi.org/10.1128/jb.96.6.2043-2048.1968

Cerofolini L, Iani C, Rosato A (2019) The quaternary structure of type II L-asparaginase from *Escherichia coli*. FEBS J 286(4):696–708

Chan WK, Lorenzi PL, Anishkin A, Purwaha P, Rogers DM, Sukharev S et al (2014) The glutaminase activity of L-asparaginase is not required for anticancer activity against ASNS-negative cells. Blood 123:3596–3606. https://doi.org/10.1182/blood-2013-10-535112

Cheng T, Cao X, Wang Y, Zhao L (2025) Development of CRISPR-Cas9-based genome editing tools for non-model microorganism *Erwinia persicina*. Synth Syst Biotechnol 10(2):555–563, ISSN 2405-805X. https://doi.org/10.1016/j.synbio.2025.02.006

Chohan SM, Rashid N (2013) TK1656, a thermostable L-asparaginase from Thermococcus kodakaraensis, exhibiting highest ever reported enzyme activity. J Biosci Bioeng 116:438–443. https://doi.org/10.1016/j.jbiosc.2013.04.005

Chow YY, Ting ASY (2015) Endophytic L-asparaginase-producing fungi from plants associated with anticancer properties. J Adv Res 6(6):869–876, ISSN 2090-1232. https://doi.org/10.1016/j.jare.2014.07.005

da Cunha MC, dos Santos Aguilar JG, de Melo RR, Nagamatsu ST, Ali F, de Castro RJS, Sato HH (2019) Fungal L-asparaginase: strategies for production and food applications. Food Res Int 126:108658, ISSN 0963-9969. https://doi.org/10.1016/j.foodres.2019.108658

Dantas, Raquel & Caetano, Ludmilla & Torres, Ariany & Alves, Matheus & Silva, Emanuelly & Teixeira, Louhanna & Teixeira, Daniel & Moreira, Renato & Fonseca, Marcela & Gaudencio Neto, Saul & Tondello Martins, Leonardo & Furtado, Gilvan & Tavares, Kaio (2019) Expression of a recombinant bacterial L-asparaginase in human cells. BMC Res Notes 12. https://doi.org/10.1186/s13104-019-4836-5

Darvishi F, Jahanafrooz Z, Mokhtarzadeh A (2022) Microbial L-asparaginase as a promising enzyme for treatment of various cancers. Appl Microbiol Biotechnol 106(17):5335–5347. https://doi.org/10.1007/s00253-022-12086-8

DeJong PJ (1972) L-asparaginase production by Streptomyces griseus. Appl Microbiol 23(6):1163–1164. https://doi.org/10.1128/am.23.6.1163-1164.1972

Deokar VD, Vetal MD, Rodrigues L (2010) Production of intracellular L-asparaginase from *Erwinia carotovora* and its statistical optimization using response surface methodology (RSM). Int J Chem Sci Appl 1(1):25–36

Deshpande N, Choubey P, Agashe M (2014) Studies on optimization of growth parameters for L-asparaginase production by *Streptomyces ginsengisoli*. Sci World J 2014:1–6. https://doi.org/10.1155/2014/895167

Dharmaraj S (2011) Study of L-asparaginase production by Streptomyces noursei MTCC 10469, isolated from marine sponge *Callyspongia diffusa*. Iran J Biotechnol 9:102–108

Distasio JA, Salazar AM, Nadji M, Durden DL (1982) Glutaminase-free asparaginase from *Vibrio succinogenes*: an antilymphoma enzyme lacking hepatotoxicity. Int J Cancer 30(3):343–347. https://doi.org/10.1002/ijc.2910300314

Domenech C, Thomas X, Chabaud S, Baruchel A, Gueyffier F, Mazingue F, Auvrignon A, Corm S, Dombret H, Chevallier P et al (2011) L-asparaginase loaded red blood cells in refractory or relapsing acute lymphoblastic leukaemia in children and adults: results of the GRASPALL 2005-01 randomized trial. Br J Haematol 153:58–65

Doriya K, Jose N, Gowda M, Kumar DS (2016) Chapter Six – Solid-state fermentation vs submerged fermentation for the production of L-asparaginase. In: Kim S-K, Toldrá F (eds) Advances in food and nutrition research, vol 78. Academic Press, pp 115–135., ISSN 1043-4526, ISBN 9780128038475. https://doi.org/10.1016/bs.afnr.2016.05.003

Ebrahiminezhad A, Rasoul-Amini S, Younes G (2011) L-asparaginase production by moderate halophilic bacteria isolated from Maharloo Salt Lake. Indian J Microbiol 51:307–311. https://doi.org/10.1007/s12088-011-0158-6

Edman P, Artursson P, Björk E, Davidsson B (1987) Immobilized L-asparaginase-L-glutaminase from *Acinetobacter glutaminasificans* in microspheres: some properties in vivo and in an extracorporeal system. Int J Pharm 34(3):225–230, ISSN 0378-5173. https://doi.org/10.1016/0378-5173(87)90184-0

Egler RA, Ahuja SP, Matloub Y (2016) L-asparaginase in the treatment of patients with acute lymphoblastic leukemia. J Pharmacol Pharmacother 7(2):62–71. https://doi.org/10.4103/0976-500X.184769

El-Naggar NE, Moawad H, El-Shweihy NM, El-Ewasy SM, Elsehemy IA, Abdelwahed NAM (2019) Process development for scale-up production of a therapeutic L-asparaginase by *Streptomyces brollosae* NEAE-115 from shake flasks to bioreactor. Sci Rep 9(1):13571. https://doi.org/10.1038/s41598-019-49709-6

El-Naggar NE, Hamouda RA, Elshafey N (2024) Artificial intelligence-based optimization for extracellular L-glutaminase free L-asparaginase production by *Streptomyces violaceoruber* under solid state fermentation conditions. Sci Rep 14(1):29625. https://doi.org/10.1038/s41598-024-77867-9

Elshafei A, El-Ghonemy D (2015) Screening and media optimization for enhancing L-asparaginase production, an anticancer agent, from different filamentous fungi in solid state fermentation. Br Biotechnol J 9:1–15. https://doi.org/10.9734/BBJ/2015/19728

Erva RR, Goswami AN, Suman P, Vedanabhatla R, Rajulapati SB (2016) Optimization of L-asparaginase production from novel *Enterobacter* sp., by submerged fermentation using response surface methodology. Prep Biochem Biotechnol 47(3):219–228. https://doi.org/10.1080/10826068.2016.1201683

Fatima N, Khan MM, Khan IA (2019) L-asparaginase produced from soil isolates of *Pseudomonas aeruginosa* shows potent anti-cancer activity on HeLa cells. Saudi J Biol Sci 26(6):1146–1153, ISSN 1319-562X. https://doi.org/10.1016/j.sjbs.2019.05.001

Ferrara MA, Severino NMB, Mansure JJ, Martins AS, Oliveira EMM, Siani AC, Pereira N, Torres FAG, Bon EPS (2006) Asparaginase production by a recombinant *Pichia pastoris* strain harbouring Saccharomyces cerevisiae ASP3 gene. Enzym Microb Technol 39(7):1457–1463, ISSN 0141-0229. https://doi.org/10.1016/j.enzmictec.2006.03.036

Florencio C, Cunha FM, Badino AC, Farinas CS, Ximenes E, Ladisch MR (2016) Secretome analysis of *Trichoderma reesei* and *Aspergillus niger* cultivated by submerged and sequential fermentation processes: enzyme production for sugarcane bagasse hydrolysis. Enzym Microb Technol 90:53–60

Ghosh S, Murthy S, Govindasamy S et al (2013) Optimization of L-asparaginase production by *Serratia marcescens* (NCIM 2919) under solid state fermentation using coconut oil cake. Sustain Chem Process 1:9. https://doi.org/10.1186/2043-7129-1-9

Gonçalves AB, Maia ACF, Rueda JA, Vanzela APDFC (2017) Fungal production of the anti-leukemic enzyme L-asparaginase: from screening to medium development. Acta Sci Biol Sci 38(387):394. https://doi.org/10.4025/actascibiolsci.v38i4.32993

Gulati R, Saxena RK, Gupta R (1997) A rapid plate assay for screening L-asparaginase producing micro-organisms. Lett Appl Microbiol 24:23–26. https://doi.org/10.1046/j.1472-765X.1997.00331.x

Hassan S, Farag A, Beltagy E (2018) Purification, characterization and anticancer activity of L-asparaginase produced by marine *Aspergillus terreus*. J Pure Appl Microbiol 12:1845–1854. https://doi.org/10.22207/JPAM.12.4.19

Hill BD, Zak A, Khera E, Wen F (2018) Engineering virus-like particles for antigen and drug delivery. Curr Protein Pept Sci 19:112–127

Honfoga JNB, Nascimento PA, Ferreira AN, da Silva EC, Bauer LC, Teixeira JM, Santana NB, Bonomo RCF (2025) Production and characterization of L-asparaginase by *Aspergillus niger*: a sustainable use of pitaya (Hylocereus spp.) by-products. Sustain Chem Environ 10:100255, ISSN 2949-8392. https://doi.org/10.1016/j.scenv.2025.100255

Isaac GS, Abu-Tahon MA (2016) Production of extracellular anti-leukemic enzyme L-asparaginase from *Fusarium solani* AUMC 8615 grown under solid-state fermentation conditions: purification and characterization of the free and immobilized enzyme. Egypt J Bot 56(3):799–816

Jain R, Zaidi KU, Verma Y, Saxena P (2012) L-asparginase: a promising enzyme for treatment of acute lymphoblastic leukemia. People J Sci Res 5(1):29–35

Jana A, Biswas S, Ghosh R, Modak R (2024) Recent advances in L-asparaginase enzyme production and formulation development for acrylamide reduction during food processing. Food Chem X 25:102055. https://doi.org/10.1016/j.fochx.2024.102055

Jayam D, Kannan G (2014) The various sources of L-asparaginase. Int J Recent Sci Res 5:342–346

Jha SK, Pasrija D, Sinha RK, Singh HR, Nigam VK, Vidyarthi AS (2012) Microbial L-asparaginase: a review on current scenario and future prospects. Int J Pharm Sci Res 3(9):3076–3090

Jha SK, Pasrija D, Singh HR, Nigam VK, Vidyarthi AS (2014) Batch production and media optimization of anti-leukemic L-asparaginase from Pseudomonas fluorescens by Taguchi DOE methodology. J Appl Pharm Sci 4(12):85–89

Jia R, Wan X, Geng X, Xue D, Xie Z, Chen C (2021) Microbial L-asparaginase for application in acrylamide mitigation from food: current research status and future perspectives. Microorganisms 9(8):1659. https://doi.org/10.3390/microorganisms9081659

Joshi RD, Kulkarni NS (2016) Optimization studies on L-asparaginase production from endophytic bacteria. Int J Appl Res 2:624–629

Kafkewitz D, Bendich A (1983) Enzyme-induced asparagine and glutamine depletion and immune system function. Am J Clin Nutr 37:1025–1030

Kamble KD, Bidwe PR, Muley VY, Kamble LH, Bhadange DG, Musaddiq M (2012) Characterization of L-asparaginase producing bacteria from water, farm and saline soil. Biosci Discov 3:116–119

Kassab M (2023) Screening and production of fungal L-asparaginase enzymes as anticancer agents from high-contrast soil environments in Egypt. Egypt Acad J Biol Sci G Microbiol 15(1):21–41. https://doi.org/10.21608/eajbsg.2023.283740

Khamna S, Yokota A, Lumyong S (2009) L-asparaginase production by actinomycetes isolated from some Thai medicinal plant rhizosphere soils. Int J Integr Biol 6(1):22–26

Kishore V, Nishita KP, Manonmani HK (2015) Cloning, expression and characterization of L-asparaginase from *Pseudomonas fluorescens* for large scale production in *E. coli* BL21. 3 Biotech 5(6):975–981. https://doi.org/10.1007/s13205-015-0300-y

Koval I, Shlyakhtina Y, Serebryakova M, Kurenkova V, Kulikov K, Shubin V et al (2021) New insights into the catalytic mechanism of type II L-asparaginases. FEBS J 288(17):5092–5107

Kumar AK, Savitha J (2015) Purification and characterization of therapeutic enzyme L-asparaginase from a tropical soil fungal isolate *Fusarium culmorum* ASP-87. MOJ Proteom Bioinform 2:1–6

Kumar DS, Sobba K (2013) L-asparaginase from microbes: a comprehensive review. Adv Biores 3:137–157

Kurinomaru T, Shiraki K (2015) Noncovalent PEGylation of L-asparaginase using PEGylated polyelectrolyte. J Pharm Sci 104:587–592

Li X, Zhang X, Xu S, Zhang H, Xu M, Yang T, Wang L, Qian H, Zhang H, Fang H, Osire T, Rao Z, Yang S (2018) Simultaneous cell disruption and semi-quantitative activity assays for high-throughput screening of thermostable L-asparaginases. Sci Rep 8:1–12. https://doi.org/10.1038/s41598-018-26241-7

Lima AM, de Paiva GB, De Souza CJ, Marques C, dos Santos JF, de Miranda RC et al (2020) Production process optimization of recombinant *Erwinia carotovora* L-asparaginase II in *Escherichia coli* fed-batch cultures and analysis of antileukemic potential. ACS Omega 9(12):1–12. https://doi.org/10.1021/acsomega.4c04711

Liu FS, Zajic JE (1973) Fermentation kinetics and continuous process of L-asparaginase production. Appl Microbiol 25(1):92–96. https://doi.org/10.1128/am.25.1.92-96.1973

Lubkowski J, Vanegas J, Chan W-K, Lorenzi PL, Weinstein JN, Sukharev S, Fushman D, Rempe S, Anishkin A, Wlodawer A (2020) Mechanism of catalysis by L-asparaginase. Biochemistry 59:1927–1945. https://doi.org/10.1021/acs.biochem.0c00116

Maggi M, Meli M, Colombo G et al (2021) Revealing *Escherichia coli* type II L-asparaginase active site flexible loop in its open, ligand-free conformation. Sci Rep 11:18885. https://doi.org/10.1038/s41598-021-98455-1

Mahajan RV, Saran S, Kameswaran K, Kumar V, Saxena RK (2012) Efficient production of L-asparaginase from Bacillus licheniformis with low-glutaminase activity: optimization, scale up and acrylamide degradation studies. Bioresour Technol 125:11–16, ISSN 0960-8524. https://doi.org/10.1016/j.biortech.2012.08.086

Martău GA, Unger P, Schneider R, Venus J, Vodnar DC, López-Gómez JP (2021) Integration of solid state and submerged fermentations for the valorization of organic municipal solid waste. J Fungi 7(9):766. https://doi.org/10.3390/jof7090766

Meena B et al (2015) L-asparaginase from Streptomyces griseus NIOT-VKMA29: optimization of process variables using factorial designs and molecular characterization of L-asparaginase gene. Sci Rep 5:1–12

Meneguetti GP, Santos JHPM, Obreque KMT, Barbosa CMV, Monteiro G et al (2019a) Correction: novel site-specific PEGylated L-asparaginase. PLoS One 14(10):e0224592. https://doi.org/10.1371/journal.pone.0224592

Meneguetti GP, Santos JHPM, Obreque KMT, Barbosa CMV, Monteiro G, Farsky SHP, Marim de Oliveira A, Angeli CB, Palmisano G, Ventura SPM, Pessoa-Junior A, de Oliveira Rangel-Yagui

C (2019b) Novel site-specific PEGylated L-asparaginase. PLoS One 14(2):e0211951. https://doi.org/10.1371/journal.pone.0211951
Michalska K, Jaskolski M, Dauter Z (2006) Structural aspects of L-asparaginases, their friends and relations. Acta Biochim Pol 53(4):627–640
Modi T, Gervais D (2022) Improved pharmacokinetic and pharmacodynamic profile of a novel PEGylated native *Erwinia chrysanthemi* L-asparaginase. Investig New Drugs 40(1):21–29. https://doi.org/10.1007/s10637-021-01173-8
Mostafa Y, Alrumman S, Alamri S, Hashem M, Al-Izran K, Alfaifi M, Elbehairi SE, Taha T (2019) Enhanced production of glutaminase-free L-asparaginase by marine *Bacillus velezensis* and cytotoxic activity against breast cancer cell lines. Electron J Biotechnol 42:6–15, ISSN 0717-3458,. https://doi.org/10.1016/j.ejbt.2019.10.001
Muneer F, Siddique MH, Azeem F, Rasul I, Muzammil S, Zubair M, Afzal M, Nadeem H (2020) Microbial L-asparaginase: purification, characterization and applications. Arch Microbiol 202(5):967–981. https://doi.org/10.1007/s00203-020-01814-1
Muslim SN (2014) Production, purification and characterization of a novel L-asparaginase from *Acinetobacter baumannii* with anti cancerous activity. Int J Curr Eng Technol 4(1):6–11
Narayana KJP, Kumar KG, Vijayalakshmi M (2008) L-asparaginase production by *Streptomyces albidoflavus*. Indian J Microbiol 48(3):331–336. https://doi.org/10.1007/s12088-008-0018-1
Naveena B, Narayani TG, Sakthiselvan P, Partha N (2012) Antioxidant and antimicrobial efficacies of Amaranthus polygonoides and its impact on L-asparaginase production. Afr J Biotechnol 11:12483–12490. https://doi.org/10.5897/ajb12.098
Nelson KE, Nelson BJ (2012) Genetic diversity of microbial endophytes and their biotechnical applications. In: Strobels G (ed) Genomics applications for the developing world, advances in microbial ecology. Springer Science + Business Media, New York, pp 249–262
Nguyen HA, Su Y, Lavie A (2016) Design and characterization of *Erwinia chrysanthemi* L-asparaginase variants with diminished L-glutaminase activity. J Biol Chem 291(34):17664–17676. https://doi.org/10.1074/jbc.M116.728485
Oza VP, Parmar PP, Patel DH, Subramanian RB (2011) Cloning, expression and characterization of L-asparaginase from *Withania somnifera* L. for large scale production. 3 Biotech 1(1):21–26. https://doi.org/10.1007/s13205-011-0003-y
Paglla U, Rao CSVR, Rajulapati SB (2013) Studies on L-asparaginase production by using *Staphylococcus capitis*. J Chem Biol Phys Sci 3(1):201–209
Pallem C (2019) Solid-state fermentation of corn husk for the synthesis of asparaginase by *Fusarium oxysporum*. Asian J Pharm Pharmacol 5(4):678–681
Parashiva J, Nuthan BR, Rakshith D, Satish S (2023) Endophytic fungi as a promising source of anticancer L-asparaginase: a review. Curr Microbiol 80(9):282. https://doi.org/10.1007/s00284-023-03392-z
Pedroso A, Miranda J, Lefin N, Effer B, Reyanldo EP, Calle Y, Monteiro G, Pessoa A Jr, Farias JG (2025) Toward safer biotherapeutics: expression and characterization of a humanized chimeric L-asparaginase in *E. coli*. Int J Mol Sci 26(14):6919. https://doi.org/10.3390/ijms26146919
Pokrywka K, Grzechowiak M, Sliwiak J, Worsztynowicz P, Loch JI, Ruszkowski M, Gilski M, Jaskolski M (2024) Probing the active site of Class 3 L-asparaginase by mutagenesis. I. Tinkering with the zinc coordination site of ReAV. Front Chem 12:1381032. https://www.frontiersin.org/journals/chemistry/articles/10.3389/fchem.2024.1381032, https://doi.org/10.3389/fchem.2024.1381032. ISSN=2296-2646
Pokrywka K, Grzechowiak M, Sliwiak J, Worsztynowicz P, Loch JI, Ruszkowski M, Gilski M, Jaskolski M (2025) Probing the active site of Class 3 L-asparaginase by mutagenesis: mutations of the Ser-Lys tandems of ReAV. Biomolecules 15(7):944. https://doi.org/10.3390/biom15070944
Pourhossein M, Korbekandi H (2014) Cloning, expression, purification and characterisation of Erwinia carotovora L-asparaginase in *Escherichia coli*. Adv Biomed Res 3:82. https://doi.org/10.4103/2277-9175.127995

Pritsa AA, Kyriakidis DA (2001a) L-asparaginase of thermus thermophilus: purification, properties and identification of essential amino acids for its catalytic activity. Mol Cell Biochem 216:93–101. https://doi.org/10.1023/A:1011066129771

Pritsa AA, Kyriakidis DA (2001b) Purification, characterization and molecular cloning of a hexameric L-asparaginase from *Thermus thermophilus*. Eur J Biochem 268(21):5483–5491

Qeshmi FI, Homaei A, Fernandes P, Javadpoure S (2018) Marine microbial L-asparaginase: biochemistry, molecular approaches and applications in tumor therapy and in food industry. Microbiol Res 208:99–112. https://doi.org/10.1016/j.micres.2018.01.011

Rabelo NG, Simões LA, Fernandes NDAT, Souza AC, Fernandes MLP, Veríssimo LAA, Schwan RF, Dias DR (2025) Optimization of L-asparaginase production from *Aspergillus caespitosus*: solid-state and submerged fermentation using low-cost substrates and partial purification. Appl Microbiol 5(1):19. https://doi.org/10.3390/applmicrobiol5010019

Rath G, Nivedita S, Behera SS, Behera HT, Gouda SK, Raina V, Achary KG, Behera SK, Ray L (2023) L-asparaginase producing novel *Streptomyces* sp. HB2AG: optimization of process parameters and whole genome sequence analysis. 3 Biotech 13(6):201. https://doi.org/10.1007/s13205-023-03620-0

Rossi L, Pierigè F, Aliano MP, Magnani M (2020) Ongoing developments and clinical progress in drug-loaded red blood cell technologies. BioDrugs 34:265–272

Rudrapati P, Audipudi AV (2017) Production and purification of anticancer enzyme L-asparaginase from Bacillus firmus AVP 18 of mangrove sample through submerged fermentation. Int J Curr Microbiol App Sci 5:1–18

Saeed H, Elsawy E, Shalaby M, Abdel-Fattah M, Hemida A, Eldoksh A, … El-Nikhely N (2021) Lasparaginase from Dickeya chrysanthemi: expression, purification and cytotoxicity assessment. Prep Biochem Biotechnol 52(6):668–680. https://doi.org/10.1080/10826068.2021.1983831

Safary A, Moniri R, Hamzeh-Mivehroud M, Dastmalchi S (2019) Highly efficient novel recombinant L-asparaginase with no glutaminase activity from a new halo-thermotolerant *Bacillus* strain. Bioimpacts 9(1):15–23. https://doi.org/10.15171/bi.2019.03

Sahu MK, Sivakumar K, Poorani E, Thangaradjou T, Kannan L (2007) Studies on L-asparaginase enzyme of actinomycetes isolated from estuarine fishes. J Environ Biol 28(2):465–474

Salzer WL, Asselin BL, Plourde PV, Corn T, Hunger SP (2014) Development of asparaginase *Erwinia chrysanthemi* for the treatment of acute lymphoblastic leukemia. Ann N Y Acad Sci 1329:81–92. https://doi.org/10.1111/nyas.12496

Samudio I, Konopleva M (2013) Asparaginase unveils glutamine-addicted AML. Blood 122(20):3398–3400. https://doi.org/10.1182/blood-2013-09-526392

Sanjotha G, Manawadi SI (2017) Isolation, screening, optimization and production of anti-tumor L-asparaginase by fungi from karwar coastal region. Res J Recent Sci 6(3):1–7

Sarquis MI, Oliveira EM, Santos AS, Costa GL (2004) Production of L-asparaginase by filamentous fungi. Mem Inst Oswaldo Cruz 99:489–492. https://doi.org/10.1590/s0074-02762004000500005

Schnuchel A, Radcke C, Theobald L, Doeding S (2023) Quality comparison of a state-of-the-art preparation of a recombinant L-asparaginase derived from Escherichia coli with an alternative asparaginase product. PLoS One 18(6):e0285948. https://doi.org/10.1371/journal.pone.0285948

Shafqat I, Shahzad S, Yasmin A, Chaudhry MT, Ahmed S, Javed A, Afzal I, Bibi M (2023) Characterization and applications of glutaminase free L-asparaginase from indigenous *Bacillus halotolerans* ASN9. PLoS One 18(11):e0288620. https://doi.org/10.1371/journal.pone.0288620

Shakambari G, Sameer Kumar R, Ashokkumar B et al (2018) Cloning and expression of L-asparaginase from *Bacillus tequilensis* PV9W and therapeutic efficacy of solid lipid particle formulations against cancer. Sci Rep 8:18013. https://doi.org/10.1038/s41598-018-36161-1

Shanmugaprakash M, Jayashree C, Vinothkumar V et al (2015) Biochemical characterization and antitumor activity of three phase partitioned L-asparaginase from *Capsicum annuum* L. Sep Purif Technol 142:258–267

Sharma D, Mishra A (2021) L-asparaginase production in solid-state fermentation using *Aspergillus niger*: process modeling by artificial neural network approach. Prep Biochem Biotechnol 52(5):549–560. https://doi.org/10.1080/10826068.2021.1972426

Shi R, Liu Y, Mu Q, Jiang Z, Yang S (2017) Biochemical characterization of a novel L-asparaginase from *Paenibacillus barengoltzii* being suitable for acrylamide reduction in potato chips and mooncakes. Int J Biol Macromol 96:93–99, ISSN 0141-8130. https://doi.org/10.1016/j.ijbiomac.2016.11.115

Shishparenok AN, Gladilina YA, Zhdanov DD (2023) Engineering and expression strategies for optimization of L-asparaginase development and production. Int J Mol Sci 24(20):15220. https://doi.org/10.3390/ijms242015220

Singh Y, Srivastava S (2013) L-asparaginase production by a new isolate Bacillus aryabhattai strain ITBHU02 in solid state culture. In: First international conference on biosciences and bioengineering: a collaborative approach.

Srikhanta YN, Atack JM, Beacham IR, Jennings MP (2013) Distinct physiological roles for the two L-asparaginase isozymes of Escherichia coli. Biochem Biophys Res Commun 436(3):362–365. https://doi.org/10.1016/j.bbrc.2013.05.066

Sudhir AP, Dave BR, Trivedi KA, Subramanian RB (2012) Production and amplification of an L-asparaginase gene from actinomycete isolate *Streptomyces* ABR2. Ann Microbiol 62:1609–1614

Sundaram SS, Kannan A, Chintaluri PG et al (2024) Thermostable bacterial L-asparaginase for polyacrylamide inhibition and *in silico* mutational analysis. Int Microbiol 27:1765–1779. https://doi.org/10.1007/s10123-024-00493-y

Swain AL, Jaskolski M, Housset D, Rao JKM, Wlodawer A (1993a) Crystal structure of Escherichia coli L-asparaginase, an enzyme used in cancer therapy (Elspar). Proc Natl Acad Sci USA 90(24):11607–11611

Swain AL, Jaskólski M, Housset D, Rao JK, Wlodawer A (1993b) Crystal structure of Escherichia coli L-asparaginase, an enzyme used in cancer therapy. Proc Natl Acad Sci USA 90(4):1474–1478. https://doi.org/10.1073/pnas.90.4.1474

Vidya J, Vasudevan UM, Soccol CR, Pandey A (2011) Cloning, functional expression and characterization of L-asparaginase II from E. coli MTCC 739. Food Technol Biotechnol 49:286–290

Wahab MRA, Palaniyandi T, Wyson J, Sivaji A, Thamada S (2024) Extracellular L-asparaginase synthesis *Bacillus niacin* isolation, optimization, and characterization from marine saltern sediment sources. Avicenna J Med Biotechnol 16(1):40–48. https://doi.org/10.18502/ajmb.v16i1.14170

Xu F, Oruna-Concha M-J, Elmore JS (2016) The use of asparaginase to reduce acrylamide levels in cooked food. Food Chem 210:163–171, ISSN 0308-8146. https://doi.org/10.1016/j.foodchem.2016.04.105

Xue D, Zhang T, Jiang B, Mu W, Zuo S (2014) Biochemical characterization of an extremely thermostable L-asparaginase from *Thermococcus gammatolerans* EJ3. J Mol Catal B Enzym 109:122–129. https://doi.org/10.1016/j.molcatb.2014.08.021

Yu M, Henning R, Walker A, Kim G, Perroy A, Alessandro R, Virador V, Kohn EC (2012) L-asparaginase inhibits invasive and angiogenic activity and induces autophagy in ovarian cancer. J Cell Mol Med 16:2369–2378. https://doi.org/10.1111/j.1582-4934.2012.01547.x

Yu W-J, Lee HY, Kwon YM, Bae SS, Choi G, Hwang H-J, Chung D (2025) The characterization of L-asparaginase with low L-glutaminase activity produced by the marine *Pseudomonas* sp. strain GH-W2b. Microbiol Res 16(1):2. https://doi.org/10.3390/microbiolres16010002

Zhang YQ, Tao ML, Shen WD, Zhou YZ, Ding Y, Ma Y, Zhou WL (2004) Immobilization of L-asparaginase on the microparticles of the natural silk sericin protein and its characters. Biomaterials 17:3751–3759

Structure and Function of Microbial Cholesterol Oxidase and Its Applications

10

Vivek Sharma, Rhydum Sharma, Randhir Kaur, and Fauziah Marpani

Abstract

Cholesterol oxidase (COX, EC 1.1.3.6), a flavoprotein alcohol oxidoreductase, plays a crucial role in cholesterol catabolism by catalyzing its two-step transformation into cholestenone and hydrogen peroxide. This monomeric enzyme's catalytic efficiency and stability are significantly influenced by its flavin adenine dinucleotide (FAD) cofactor and interactions with lipid bilayers. Microbial COX, with its unique structure and mechanism, offers broad biotechnological applications, extending beyond its physiological significance. It is widely utilized in clinical diagnostics for serum cholesterol determination, biocatalysis of steroid intermediates, as an insecticidal agent in transgenic crops, and as a molecular probe for cellular membrane studies. This chapter comprehensively reviews the diverse microbial sources of cholesterol oxidase, methods for its purification and characterization, and its extensive applications, particularly within the pharmaceutical industry.

Keywords

Sterol · Indicator · Health · Medicine

V. Sharma (✉) · R. Sharma · R. Kaur
University Centre for Research and Development, Chandigarh University, Mohali, Punjab, India

F. Marpani (✉)
Faculty of Chemical Engineering, Universiti Teknologi MARA, 40450 Shah Alam, Selangor, Malaysia
e-mail: fauziah176@uitm.edu.my

© The Author(s), under exclusive license to Springer Nature Singapore Pte Ltd. 2026
V. Sharma, R. Salwan (eds.), *Pharmaceutically Relevant Microbial Enzymes*,
https://doi.org/10.1007/978-981-95-4941-2_10

10.1 Introduction

Cholesterol is one of the vital constituents of the eukaryotic membranes due to its role in the cellular integrity, cell signaling, and precursor for the biosynthesis of steroid hormones and bile acids (Russell 2003). The dysregulation of cholesterol metabolism can cause various pathological disorders (Mahley and Bersot 2017). For example, the accumulation of cholesterol in plasma often causes the development of diseases such as atherosclerosis or coronary heart diseases (Rose and Shipley 1980; MacLachlan et al. 2000). Hence, the cholesterol metabolism is of utmost importance in clinical diagnostics. Cholesterol oxidase (EC 1.1.3.6) is a monomeric oxidoreductase flavoenzyme. The enzyme responsible for the catabolism of sterols catalyzes the oxidation of cholesterol to cholesterone and hydrogen peroxide (Shao et al. 2019). The enzyme is also used in combination with other enzymes for the clinical determination of serum cholesterol levels (Flegg 1973). The other applications of COX include the transformation of steroids and nonsteroids (Richmond 1973), insecticidal agent for pest management in transgenic crops (Richmond 1973; Purcell et al. 1993; Wei et al. 2012), lowering the cholesterol level in foods (Rose and Shipley 1980), as biosensors for cholesterol detection, and its cellular interactions and membrane dynamics (MacLachlan et al. 2000; Suckling and Suckling 1994).

Cholesterol oxidase assist in the microbial-mediated catabolism of cholesterol (Ferrer et al. 1996). So far, COX has been reported from different members of the genus Actinomycetes, particularly from the genus such as *Rhodococcus*, *Streptomyces*, and *Nocardia* has been extensively explored (Torigoe et al. 1999; Luo et al. 2011). Due to their innate capabilities to thrive in cholesterol-rich substrates, these microorganisms are a good source of enzyme and their encoding genes. Furthermore, the genetic and biochemical diversity of COX from microbial sources such as pH and temperature optima, thermal stability, and substrate specificity are often explored for various biotechnological applications (Yamada and Kondo 1982). Due to the catalytic efficiency and specificity, microbial cholesterol oxidases have been used for plethora of applications described earlier. One of the most widely recognized of COX usage involves the enzymatic estimation of cholesterol and high-density lipoprotein (HDL)-conjugated cholesterol in serum and plasma (Allain et al. 1974). Here, the coupling with peroxidase is used for rapid detection of hypercholesterolemia and cardiovascular risk. Additionally, COX is also explored for the regioselective transformation of wide range of sterols and their intermediates in the pharmaceutical industries which are otherwise difficult to synthesize (Schmid et al. 2001).

Time to time efforts have been made on the exploration of COX from different microbial sources and on improving their performance. Using genetic engineering and directed evolution approaches, the COX mutant with higher thermal stability, improved substrate specificity, and a broad pH and temperature optima has been developed to meet the industrial demands (Choi et al. 2005; Du et al. 2010). Parallel efforts have been made on enhancing the reusability and stability by immobilization of enzyme to reduce the cost for industrial processes (Mohy-Eldin et al. 2016).

Efforts also have been made for the exploration of novel COX from different microorganisms including extremophiles such as thermophiles and halophiles (Sharma et al. 2018).

Keeping in view the importance of microbial COX, this chapter describes the microbial cholesterol oxidase and their applications in pharmaceutical applications.

10.2 Microbial Sources of Cholesterol Oxidase

Cholesterol oxidases catalyze the oxidation of cholesterol using oxygen as an electron acceptor in presence of flavin adenine dinucleotide (FAD) (Kojima et al. 2013). Cholesterol oxidases (COX) are flavoenzymes that catalyze the oxidation and isomerization using one active site (Moradpour and Ghasemian 2016). This enzyme has been extensively explored in sectors such as agriculture, biotechnology, chemistry, medicine, and pharmacy due to its capacity to oxidize recalcitrant substrate such as 3β-hydroxy-5-ene- to 3-keto-4-ene- steroids and formation of hydrogen peroxide as side product (Fokina et al. 2022) (Fig. 10.1). In clinical laboratories, this enzyme has been extensively used for cholesterol detection. The potential of bacteria to catabolize the cholesterol is known for several years. For example, in 1947, Turfitt showed that *Proactinomyces erythropolis* now *Nocardia erythropolis* was responsible for the oxidation of cholesterol into 4-cholesten-3-one from cholesterol (Turfitt 1947). Then in 1949, a soil mycobacterium was reported for the oxidation for cholesterol (Schatz et al. 1949). In later study, this bacterium was investigated for the formation of 4-cholesten-3-one when they incubated cholesterol with the cell-free extract of bacteria. It was observed that the dialysis of filtrate led to loss of enzyme activity (Stadtman et al. 1954; MacLachlan et al. 2000). In subsequent study, it was reported that the enzyme acted as a true NAD-dependent dehydrogenase, whereas others catalyze the oxidations of sterols and only require O_2 for its activity (Smith and Brooks 1976). In 1972–1973, the enzyme responsible for cholesterol metabolism was identified as cholesterol oxidase (COX) and used for serum cholesterol estimation (Richmond 1973; Flegg 1973). Flegg identified that *Nocardia erythropolis* produced the COX in the presence of cholesterol as the only carbon

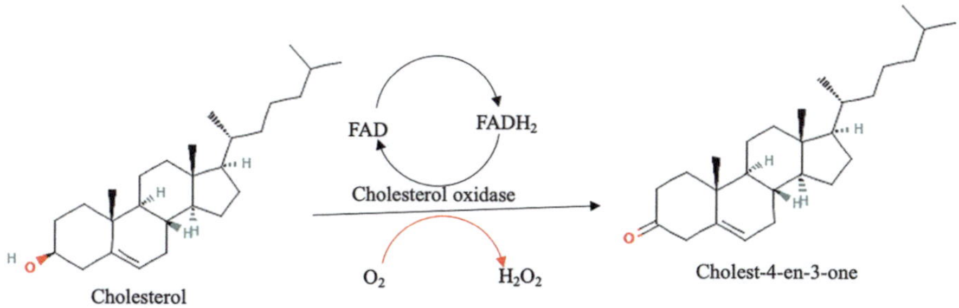

Fig. 10.1 Mechanism of cholesterol oxidase. Cholesterol oxidase (COX, EC 1.1.3.6), a flavoprotein alcohol oxidoreductase, plays a crucial role in cholesterol catabolism by catalyzing its two-step transformation into cholestenone and hydrogen peroxide

source. In 1973, another species of *Nocardia* (NCIB 10554) was used for the purification of enzyme (Richmond 1973; Fukuda et al. 1973). The purified COX was produced oxidase 1 mole of hydrogen peroxide (H_2O_2) per mol of cholesterol oxidized. The release of H_2O_2 was used the estimation of cholesterol in serum Now different analytical procedures have been developed and optimized for its commercial applications (Smith and Brooks 1976).

Steroids are known for their recalcitrance nature. In majority of the studies, the steroid degradation has been studied under aerobic conditions only, whereas the efforts for anaerobic degradation of steroids are restricted only. In recent study, the metagenomics-based approach has been deployed to understand the biodegradation of cholesterol via sludge microbial communities during wastewater treatment. The genomic and transcriptomic analysis revealed unique cholesterol uptake, and metabolism mechanism in β-proteobacteria (Wei et al. 2018). So far, COX has been isolated and characterized from different types of microorganisms (Moradpour and Ghasemian 2016). Members of the genus *Rhodococcus* sp., *Rhodococcus*sp. PKPD-CL (Kasabe et al. 2015) *Streptomyces* sp. (Yazdi et al. 2001; Alapati 2020) *Nocardia, Pseudomonas aeruginosa* (Doukyu and Nihei 2015), *Arthrobacter* sp., *Mycobacterium* sp., and *Streptoverticillium* (Yazdi et al. 2008; Bholay et al. 2013), *Bordetella* species (Lin et al. 2010), *Rhodococcus* sp. PKPD-CL (Kasabe et al. 2015) (Fig. 10.2) have been explored for the cholesterol oxidase production. Even COX from pathogenic bacteria has been explored for their role in affecting the cell membrane integrity as part of pathogenesis (Brzostek et al. 2007; Elsayed and Ahmed Abdelwahed 2020). The cholesterol oxidase has been reported from then genus belonging to Actinobacteria such as *Rhodococcus*, *Streptomyces*, *Mycobacterium. Rhodococcus jostii* oxidize C26 of the sterol side to stenone or sterol-C26-oic acid which is then catabolized after the cleavage of the sterol side chain and metabolism of the steroid nucleus (Kreit 2017).

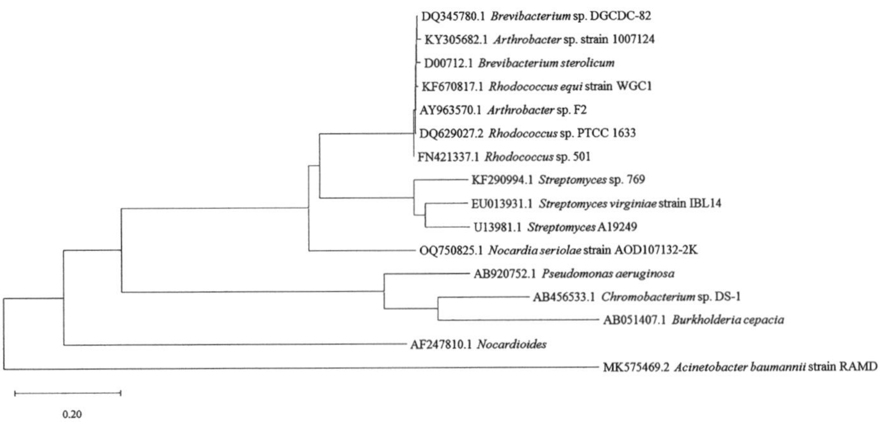

Fig. 10.2 Phylogenetic status microbial cholesterol oxidases

10.3 Purification of Cholesterol Oxidase

The molecular mass of cholesterol oxidase from in range of 40–60 kDa in size has been reported. For instance, cholesterol oxidase from *Bordetella* strain B4 (code, CGMCC 2229) obtained from soil has been reported in the range of 55 kDa in size has been determined (Lin et al. 2010), whereas almost similar size of recombinant extracellular cholesterol oxidase, i.e., 55.6 kDa in size, has been from the actinobacterium *Nocardioides simplex* VKM Ac-2033D (Fokina et al. 2022). In another study, a cholesterol oxidase of 42.84 kDa in size was purified from a marine *Streptomyces* sp. (Alapati 2020). The purification of cholesterol oxidase (ChOx) has been approached using various techniques, often tailored to the source organism and desired purity level. For instance, cholesterol oxidase has been extracted from cells using detergents like Triton X-100 (Smith and Brooks 1976). The addition of cholesterol as an emulsion in Tween 20 significantly enhances the enzyme yield. The pH and temperature optima of the enzyme was 9 and 60 °C, respectively, and the presence of metal ion such as Mg^{2+} and Zn^{2+} is known to enhance the enzyme activity (Alapati 2020).

The purification of COX has been done using various chromatographic methods such as anion-exchange chromatography on DEAE-cellulose. After washing, the elution is typically performed with a gradient of TRIS buffer (pH 8.0) or a stepwise elution with potassium phosphate buffer (pH 7.5) in presence of Triton X-100 (Smith and Brooks 1976). The ammonium sulfate precipitation for partial purification, deployed for extracellular enzymes, has been used for *Streptomyces hygroscopicus* H 82 N-SY 7. The use of polyethylene glycol 4000 precipitation has also been utilized, along with diethylaminoethyl Sephacel anionic column chromatography and Superdex-200 gel filtration, to purify cholesterol oxidase from a marine *Streptomyces* sp. (Alapati 2020). For the purification of recombinant COX, affinity chromatography such as His-tags has been reported as highly effective method, for the purification of ChONs from *Nocardioides simplex* VKM Ac-2033D (Fokina et al. 2022). Furthermore, cholesterol oxidase and cholesterol esterase from *Pseudomonas fluorescens* have been effectively adsorbed on an Amberlite CG-50 column (Sasaki et al. 1982).

10.4 Characterization and Production of Cholesterol Oxidase Gene

Due to the high cost of COX which is primarily due to low enzyme yield from microbial sources, efforts have been made to improve yield and its biochemical properties (MacLachlan et al. 2000). In early, efforts were focused on the optimization of growth conditions. For instance, the presence of oleic acid and increased agitation rate enhanced the production of cholesterol oxidase by *Schizophyllum commune* (MacLachlan et al. 2000). Additionally, the genetic engineering has been used to enhance the enzyme production and overcome the limitations of from natural sources (Moradpour and Ghasemian 2016). The cholesterol oxidase gene (*choA*) from *Streptomyces* sp. SA-COO has been overexpressed into *Streptomyces lividans* using the vector pIJ702. Similarly, the cholesterol oxidase gene (*choB*) from *Brevibacterium sterolicum* has also been cloned (Chen et al. 2006). The gene (*choAA*) encoding cholesterol oxidase from *Arthrobacter simplex* F2 was cloned and sequenced revealing a gene of 1653 bp in size encoded a protein of 551 amino acids. N-terminal sequencing confirmed that the mature enzyme consists of 502 amino acids with a predicted molecular mass of 54.2 kDa in size and a signal peptide of 49-amino acid residues. This gene was successfully cloned and efficiently expressed in *Escherichia coli* and *Pichia pastoris* (Chen et al. 2006).

The COX gene from *Pseudomonas aeruginosa* strain PA157 has been characterized for its ability to produce 6β-hydroperoxycholest-4-en-3-one-forming cholesterol oxidase, whereas the gene from *P. aeruginosa* strain PAO1 has been expressed in *Escherichia coli*. The purified enzyme was active over a wide pH range, i.e., 5.5–10, and retained ~90% of its activity after 30 min at 70 °C (Doukyu and Nihei 2015). Another cholesterol oxidase (COD) gene from *Brevibacterium* sp. (DQ345780) has been expressed in *E. coli* BL21 (DE3) and purified using riboflavin linked to Sepharose 4B. The recombinant enzyme is ~50 kDa in size with a dingle peptide (Xin et al. 2011). In 2015, a novel COX from *Rhodococcus* sp. PKPD-CL of 60 kDa in size from Chilika Lake has been characterized. The enzyme exhibited halotolerant, detergent, and organic solvent-stable alkaline cholesterol oxidase with pH and temperature optima of pH 8.0 and 37 °C, respectively (Kasabe et al. 2015). In another study, the extracellular cholesterol oxidase of 55.6 kDa in size was cloned from *Nocardioides simplex* VKM Ac-2033D and then overexpressed in recombinant protein. This recombinant enzyme exhibited promising activities against various 3β-hydroxysteroids, with a pH optima pH of 6.0, which was scomparable with commercial *Streptomyces hygroscopicus* ChO (Fokina et al. 2022).

Alternatively, an integrative approach using evolutionary analysis, classic enzymatic techniques and computational methods of various ChOx enzymes from diverse sources including *Rhodococcus* sp. (RCO), *Chromobacterium* sp. (CCO), *Pseudomonas aeruginosa* (PCO), and *Burkholderia cepacia* (BCO) has been used. The comparative analysis of the substrate-binding domain and FAD-binding domain revealed a common ancestor for all COX and with structural variations during divergence. Overall, the variation was observed for substrate-binding domain than the FAD-binding domain, irrespective of the FAD coupling mechanism (Shapira et al. 2024).

The thermal stability activity of cholesterol oxidase has been improved using random mutagenesis. For instance, the random mutagenesis of *Streptomyces* sp. SA-COO cholesterol oxidase has been used to obtain two thermostable mutants (*ChOA*-M1 and *ChOA*-M2). The structural elucidation revealed that substitutions of three amino acid residues (G49T, W52K, and F62V) for *ChOA*-M1 mutant, which enhanced the thermostability up to 40% improvement at 50 °C for 5 h as well as turnover of the enzyme. The mutagenesis also broadened pH range which is helpful for various clinical and industrial applications (Fazaeli et al. 2022). Researchers also identified new variants of ChOx; CHO-A and CHO-U. Although the N-terminal sequences of both were nearly identical, CHO-U exhibited distinct characteristics, including stability in nonionic detergents, performed well under acidic optimum pH, and also displayed better stability at alkaline pH along with a lower Km for cholesterol and a higher Vmax compared to CHO-A (Isobe et al. 2003a, b).

Among various microbial strains, *Escherichia fergusonii* has been identified one of the prolific producers of ChOx, an extracellular enzyme activity. The optimal enzyme was reported in a basal medium pH 6.5 at 48 h of incubation with shaking of 130 rpm at 35 °C, supplemented with 1.4 g/L cholesterol as the sole carbon source (Khataee et al. 2025). The commercial-scale production of cholesterol oxidase has been carried out using submerged fermentation systems (Elsayed and Ahmed Abdelwahed 2020). It was observed that enzyme yield was affected by two primary conditions: (i) optimization of the culture medium and (ii) refinement of process parameters. Further, the medium optimization can be done using either the traditional one-factor-at-a-time approach or using an advanced statistical approach. The first approach is straightforward as it is labor-intensive but incurs the higher experimental costs (Elsayed and Ahmed Abdelwahed 2020). The second approach that uses the statistical methods including factorial design and response surface methodology offers significant advantages. Here various parameters such as key nutrients and experimental conditions can be optimized with less costs and time (Elsayed and Ahmed Abdelwahed 2020).

In one study, the cholesterol oxidase has been expressed in probiotic strains: *Lactobacillus plantarum* and *Propionibacterium freudenreichii*. Here, a modified *choA* gene and its specific promoter elements from *Streptomyces* sp. were cloned. In *L. plantarum*, enzyme yield of 3.6 mU/mg was observed, whereas in *P. freudenreichii* subsp. *shermanii* IFO12426, the enzyme yield of 1.6, 4.3, and 7.2 U/mg protein under three promoters P4, P8, and P138, respectively, was achieved. The study indicated that the cholesterol oxidase producing probiotic strains (Kiatpapan et al. 2001). A cholesterol oxidase of 58 KDa in size from the γ-subclass of *Proteobacteria* from strain Y-134 has been isolated and characterized. It was also observed that the addition of 0.4% Triton X-100 helped to enhance the enzyme production at 30 °C. The purification of enzyme resulted in a protein of 115 kDa in size, due to two identical 58 kDa in size. The enzyme efficiently oxidized the cholesterol and dehydrocholesterol and displayed Km of 65 μM for cholesterol. The protein sequence analysis revealed that the enzyme shared similarity to cholesterol oxidase from *Burkholderia cepacia*, then *Streptomyces* or *Brevibacterium* (Isobe et al. 2003a, b).

10.5 Application of Cholesterol Oxidase

The cholesterol oxidases play crucial role in microbial catalyzed sterol catabolism. For this, COX catalyzes the first step in cholesterol biodegradation by oxidizing it to cholest-4-en-3-one (Shapira et al. 2024; Chernov et al. 2017). This enzyme is explored for its diverse applications. Two COX referred as ChoM1 and ChoM2 have been identified in *Mycobacterium neoaurum*. It was observed that with ChoM2 shared about 45% identity with other microbial cholesterol oxidases, with ChoM2 sharing up to 45% identity with other cholesterol oxidases. Efforts have been made to enhance the activity of cholesterol oxidases such as ChoM2 for upscaling the production of sterols to valuable steroids having potential applications in industrial sectors (Yao et al. 2013). Cholesterol oxidases have also been explored as enzyme-based sensors. Site-directed mutagenesis approach has been used by the researcher to target the oxygen-binding residues responsible for oxidase activity. For instance, the substitution of Val at 191 with Al significantly reduced the oxidase activity and enhanced the dehydrogenase activity of cholesterol oxidase that resulted in 400-fold upregulation of dehydrogenase/oxidase ratio. Therefore, such strategies are attractive method for engineering enzymes with better electrochemical cholesterol monitoring properties (Kojima et al. 2013).

10.5.1 Biotechnological and Industrial Applications

Cholesterol oxidases are one of the crucial molecular tools used in clinical diagnostics, the production of steroid drugs, and insecticides (Chernov et al. 2017). The enzyme is also used for tracking intracellular cholesterol, although its efficacy is restricted by its cost and lack of efficient temporal control over its activity (Chernov et al. 2017). Using immobilization, the stability and reusability has been improved. The cholesterol oxidases from *Pseudomonas aeruginosa* PseA, *Rhodococcus erythropolis* MTCC 3951, and *Streptomyces* sp. have been successfully immobilized using nanomaterials. The nano-formulations led to increase in catalytic efficiency, thermal stability, and also decrease in the cost of enzyme (Ghosh et al. 2018).

Similarly, ChOx has been immobilized on Layered Double Hydroxide (LDH) nanomaterials for biosensor applications, demonstrating retained activity after several uses (Temani et al. 2014). Furthermore, a nano-porous metal-organic-framework (MOF) has been used as a host for ChOx, improving its stability and catalytic efficiency for substrate oxidation (Hassanzadeh et al. 2018). The interaction of peripheral membrane protein, cholesterol oxidase, and membrane has been investigated using cholesterol, cholesterol: sphingomyelin in 1:1 ratio, and the raft model composed of dioleoylphosphatidylcholine (DOPC), cholesterol (Chol), and sphingomyelin (SM) (1:1:1) to decrease the cholesterol in hypercholesterolemia patient (Zaborowska-Mazurkiewicz et al. 2025).

10.5.2 Clinical and Diagnostic Applications

Cholesterol is a crucial biomarker for various diseases; therefore, the accurate estimation of cholesterol in human serum is vital for diagnosis and prevention (Wang et al. 2021). The cholesterol oxidases are already in use for cholesterol detection (Shapira et al. 2024; Lashkarian et al. 2010). The paper-based sensor has been designed for cholesterol detection and has been developed using screen-printed electrode (SPE) and nanoporous gold. Further, to attain the sensitivity and accuracy of cholesterol detection, a synergistic approach uses the oxidation of cholesterol by cholesterol oxidase and the reduction of H_2O_2 using nanoporous gold. The resulting electrochemical paper sensor exhibited a wider linear response (50 μM to 6 mM) and also exhibited a better sensitivity (32.68 μA mM^{-1} cm^{-2}) and a lower detection limit (8.36 μM). This cholesterol detection method such as a paper sensor using a screen-printed electrode modified with nanoporous gold leverages the oxidation of cholesterol by ChOx and the subsequent reduction of the oxidation product (H2O2) by nanoporous gold offers high sensitivity and a wide linear response range for cholesterol detection (Wang et al. 2021). Yeast surface display systems have also been utilized for electrochemical detection of cholesterol, demonstrating good performance and stability (Zhao et al. 2020). Specific ChOx variants, like those from *Streptomyces* and *Brevibacterium*, have been compared for their sensitivity in total serum cholesterol assays, with the *Streptomyces* enzyme showing higher sensitivity (Srisawasdi et al. 2006). The comparative evaluation of *Streptomyces* and *Brevibacterium* cholesterol oxidase for total serum cholesterol detection revealed the detection limit of 20.7 and 2.6 mmol/L, respectively (Srisawasdi et al. 2006). In 2020, a cholesterol oxidase from *Chromobacterium* sp. DS-1 displayed on the surface of *Saccharomyces cerevisiae* has been used detection of cholesterol and glucose. The cholesterol biosensor showed a linear range from 2 to 6 mmol/L along with glucose detection in the range of 1.4–33.3 mmol/L. Both the biosensors were stable for more than 21 days (Zhao et al. 2020).

10.6 Limitations in Membrane Probing and Novel Functions

Despite its widespread use, cholesterol oxidase has limitations in probing the distribution of cholesterol in biological membranes. Cholesterol in intact brush-border membrane vesicles is a poor substrate for cholesterol oxidases. Hence, a deeper understanding of the factors controlling the enzyme-substrate interaction is needed for cholesterol oxidases to be effectively used as a probe for cholesterol distribution in biological membranes (Thurnhofer et al. 1986). Interestingly, recent research has broadened the understanding of cholesterol oxidases role for biological functions beyond cholesterol degradation. The gene *pimE*, encoding a cholesterol oxidase in *Streptomyces natalensis*, has been used for the biosynthesis of the polyene macrolide pimaricin, an antibiotic. The extracellular cholesterol oxidase restored the pimaricin production in mutant cultures, indicating its role as signaling proteins for polyene biosynthesis (Mendes et al. 2007).

10.7 Scope of Genetic Engineering

Ongoing research focuses on improving ChOx's utility, such as developing regulatable systems for temporal control of its activity. A regulatable fragment complementation system for ChOx from *Chromobacterium* sp. has been developed, where the enzyme is split into two moieties. The addition of rapamycin reconstitutes a fluorescent and active enzyme, allowing for rapid, controlled induction of cholesterol oxidation and studies on cholesterol's impact on cellular processes (Chernov et al. 2017). Different bacterial strains, including *S. aegyptia* strain NEAE-102 and a γ-subclass Proteobacteria strain Y-134, have been identified for their high production of extracellular cholesterol oxidase, with unique characteristics such as stability in detergents (El-Naggar et al. 2018; Isobe et al. 2003a, b). Efforts are also underway to express heterologous cholesterol oxidases genes in probiotic strains like *Lactobacillus plantarum* and *Propionibacterium freudenreichii* to potentially produce useful proteins in these economically important bacteria (Kiatpapan et al. 2001). Furthermore, studies are investigating the effects of cholesterol oxidases on model lipid membranes (composed of cholesterol, cholesterol: sphingomyelin, and raft models) to understand the interaction between the enzyme and cell membranes. This research aims to find alternative strategies to decrease excessive cholesterol levels, potentially offering new approaches beyond traditional statin therapy for hypercholesterolemia (Zaborowska-Mazurkiewicz et al. 2025).

10.8 Immobilization of Cholesterol Oxidase

Microbial cholesterol oxidases have been used extensively in various fields because of their better catalytic activity. For this, the immobilization on different support materials has been used to improve the stability, reusability, and catalytic efficiency compared to its free form. For instance, cholesterol oxidases enzymes isolated from *Pseudomonas aeruginosa PseA*, *Rhodococcus erythropolis MTCC 3951*, and *Streptomyces* sp. have been successfully covalently coupled onto materials such as silane-modified iron (II, III) oxide magnetic nanoparticles. These magnetic nanoparticles-conjugated enzyme has been demonstrated remarkable for its activity retention as cholesterol oxidases retained the highest activity and with better catalytic efficiency compared to the soluble enzymes. Furthermore, the immobilized enzyme exhibited better pH, thermal stability, as it retained 50–52% of residual activity after repeated usage for ten operational cycles. The reusability of the immobilized enzyme is helpful for industrial and clinical applications (Ghosh et al. 2018). Beside magnetic nanoparticles, cholesterol oxidase has also been immobilized using materials such as "Layered Double Hydroxide" (Temani et al. 2014). The immobilized enzymes on nanomaterials exhibited two folds' activity compared to soluble enzyme (Ghosh et al. 2018). Alternatively, the nano-porous metal-organic-framework with significant improved stability and catalytic efficiency has been explored for the immobilization for cholesterol oxidases. The cholesterol oxidases and metal-organic-framework hybrid led to the development of highly sensitive and selective fluorometric cholesterol probes (Hassanzadeh et al. 2018). The diverse immobilization strategies underscore the ongoing advancements in making ChOx a more robust and versatile enzyme for a wide range of applications.

10.9 Conclusion and Future Perspectives

Microbial cholesterol oxidase (COX) stands out as a highly versatile flavoenzyme with profound implications across clinical, industrial, and agricultural sectors. Its capacity to oxidize cholesterol, a critical step in sterol catabolism, underpins its various applications, from precise cholesterol quantification in medical diagnostics to the biocatalytic synthesis of valuable steroid intermediates. *Rhodococcus*, *Streptomyces*, *Nocardia*, *Pseudomonas*, and *Mycobacterium* showcase their unique biochemical properties such as pH and temperature optima, and substrate specificities. Significant advancements in genetic engineering and directed evolution have yielded COX mutants with enhanced thermal stability, improved substrate specificity, and broader pH and temperature ranges, meeting industrial demands and paving the way for more efficient and cost-effective processes. Furthermore, immobilization techniques have markedly improved the enzyme's reusability and stability, crucial for large-scale industrial applications and the development of robust biosensors. The continuous exploration and characterization of novel COX enzymes from extremophiles and other microorganisms further expand their biotechnological potential.

The future of microbial cholesterol oxidase research is rich with potential, suggesting several promising avenues for exploration. Continued bioprospecting for novel COX enzymes from diverse and extreme microbial environments holds the key to uncover variants with superior catalytic properties, enhanced stability under harsh industrial conditions, and broader substrate specificities. Furthermore, the application of advanced enzyme engineering strategies, including rational design, directed evolution, and computational approaches, will be crucial for tailoring COX to meet specific industrial and clinical demands, such as improving its efficiency in non-aqueous systems or enhancing its tolerance to inhibitors. Significant efforts will also be directed towards developing cost-effective and scalable production methods for recombinant COX, through optimized microbial fermentation and purification protocols, which are essential for its widespread commercial adoption. The development of integrated bioprocesses utilizing immobilized or engineered COX for the efficient production of high-value steroid precursors and pharmaceuticals remains a pivotal area for innovation. Additionally, exploring the potential of COX in novel biosensor designs for rapid, accurate, and portable cholesterol detection, particularly in point-of-care settings, could revolutionize the clinical diagnostics. Finally, a deeper understanding of the structural and mechanistic aspects of COX-lipid interactions promises to lead to novel therapeutic strategies aimed at cholesterol metabolism or disrupting bacterial pathogenesis.

References

Alapati K (2020) Characterization of cholesterol oxidase from a marine Streptomyces sp. and its cytotoxicity. Process Biochem 89:175–185

Allain CC, Poon LS, Chan CS, Richmond W, Fu PC (1974) Enzymatic determination of total serum cholesterol. Clin Chem 20(4):470–475

Bholay BD, Gadekar DJ, Sawant SK, Sonawane SM (2013) Bacterial extracellular cholesterol oxidase and its pharmaceutical perspectives. Int J Curr Microbiol Appl Sci 2:19–28

Brzostek A, Dziadek B, Rumijowska-Galewicz A, Pawelczyk J, Dziadek J (2007) Cholesterol oxidase is required for virulence of Mycobacterium tuberculosis. FEMS Microbiol Lett 275(1):106–112

Chen YR, Huang HH, Cheng YF, Tang TY, Liu WH (2006) Expression of a cholesterol oxidase gene from Arthrobacter simplex in Escherichia coli and Pichia pastoris. Enzym Microb Technol 39(4):854–860

Chernov KG, Neuvonen M, Brock I, Ikonen E, Verkhusha VV (2017) Introducing inducible fluorescent split cholesterol oxidase to mammalian cells. J Biol Chem 292(21):8811–8822

Choi HS, Kim YS, Shin HJ, Kwak MS, Lee MS, Kim MK (2005) Enhancing the thermostability of cholesterol oxidase from *Streptomyces lividans* by directed evolution. Enzym Microb Technol 37(1):143–149

Doukyu N, Nihei S (2015) Cholesterol oxidase with high catalytic activity from Pseudomonas aeruginosa: screening, molecular genetic analysis, expression and characterization. J Biosci Bioeng 120(1):24–30

Du JB, Deng Y, Han RM, Zhang YZ (2010) Directed evolution of cholesterol oxidase from *Arthrobacter simplex* for improved activity and stability in organic solvents. J Mol Catal B Enzym 67(3–4):213–219

El-Naggar NE, Soliman HM, El-Shweihy NM (2018) Extracellular cholesterol oxidase production by Streptomyces aegyptia, in vitro anticancer activities against rhabdomyosarcoma, breast cancer cell-lines and in vivo apoptosis. Sci Rep 8(1):2706. https://doi.org/10.1038/s41598-018-20786-3. PMID: 29426900; PMCID: PMC5807524

Elsayed EA, Ahmed Abdelwahed N (2020) Medium optimization by response surface methodology for improved cholesterol oxidase production by a newly isolated *Streptomyces rochei* NAM-19 strain. Biomed Res Int 2020:1870807. https://doi.org/10.1155/2020/1870807. PMID: 32337225; PMCID: PMC7150710

Fazaeli A, Fana SE, Golestani A, Aminian M (2022) Improvement of thermostability of cholesterol oxidase from streptomyces Sp. SA-COO by random mutagenesis. Protein Expr Purif 191:106028

Ferrer M, Nielsen P, Jørgensen PL (1996) Purification and characterization of a cholesterol oxidase from *Streptomyces lavendulae*. Enzym Microb Technol 18(4–5):374–380

Flegg HM (1973) Ames award lecture 1972. An investigation of the determination of serum cholesterol by an enzymatic method. Ann Clin Biochem 10(1–6):79–84

Fokina VV, Karpov MV, Kollerov VV, Bragin EY, Epiktetov DO, Sviridov AV, Kazantsev AV, Shutov AA, Donova MV (2022) Recombinant extracellular cholesterol oxidase from Nocardioides simplex. Biochemistry (Mosc) 87(9):903–915. https://doi.org/10.1134/S0006297922090048. PMID: 36180991

Fukuda H, Kawakami Y, Nakamura S (1973) A method to screen anticholesterol substances produced by microbes and a new cholesterol oxidase produced by Streptomyces violascens. Pharm Bull 21:2057–2060

Ghosh S, Ahmad R, Gautam VK, Khare SK (2018) Cholesterol-oxidase-magnetic nanobioconjugates for the production of 4-cholesten-3-one and 4-cholesten-3,7-dione. Bioresour Technol 254:91–96

Hassanzadeh J, Khataee A, Eskandari H (2018) Encapsulated cholesterol oxidase in metal-organic framework and biomimetic Ag nanocluster decorated MoS2 nanosheets for sensitive detection of cholesterol. Sensors Actuators B Chem 259:402–410

Isobe K, Mori N, Wakao N (2003a) The second cholesterol oxidase produced by γ-proteobacterium Y 134. J Biosci Bioeng 96(3):257–261

Isobe K, Shoji K, Nakanishi Y, Yokoe M, Wakao N (2003b) Purification and some properties of cholesterol oxidase stable in detergents from γ-proteobacterium Y-134. J Biosci Bioeng 95(3):257–263

Kasabe PJ, Mali GT, Dandge PB (2015) Assessment of alkaline cholesterol oxidase purified from Rhodococcus sp. PKPD-CL for its halo tolerance, detergent and organic solvent stability. Protein Expr Purif 116:30–41

Khataee S, Dehghan G, Rashtbari S, Marefat A, Jamei S, Farzi-Khajeh H (2025) Optimized extraction and kinetic study of cholesterol oxidase from a newly isolated Escherichia fergusonii strain from local whey samples: insights through a combined experimental study and artificial neural network modeling. BMC Microbiol 25(1):32. https://doi.org/10.1186/s12866-024-03728-0. PMID: 39833667; PMCID: PMC11744973

Kiatpapan P, Yamashita M, Kawaraichi N, Yasuda T, Murooka Y (2001) Heterologous expression of a gene encoding cholesterol oxidase in probiotic strains of Lactobacillus plantarum and Propionibacterium freudenreichii under the control of native promoters. J Biosci Bioeng 92(5):459–465

Kojima K, Kobayashi T, Tsugawa W, Ferri S, Sode K (2013) Mutational analysis of the oxygen-binding site of cholesterol oxidase and its impact on dye-mediated dehydrogenase activity. J Mol Catal B Enzym 88:41–46

Kreit J (2017) Microbial catabolism of sterols: focus on the enzymes that transform the sterol 3β-hydroxy-5-en into 3-keto-4-en. FEMS Microbiol Lett 364(3):fnx007. https://doi.org/10.1093/femsle/fnx007. PMID: 28087615

Lashkarian H, Raheb J, Shahzamani K, Shahbani H, Shamsara M (2010) Extracellular cholesterol oxidase from Rhodococcus sp.: isolation and molecular characterization. Iran Biomed J 14(1–2):49–57. PMID: 20683498; PMCID: PMC3878146

Lin Y, Fu J, Song X (2010) Purification and characterization of an extracellular cholesterol oxidase from a Bordetella species. Process Biochem 45(9):1563–1569

Luo C, Liu C, Gu J, Liu H, Ma Y (2011) Cloning, expression, and characterization of a novel cholesterol oxidase from *Rhodococcus erythropolis* LQ1. Appl Biochem Biotechnol 164(4):488–498

MacLachlan JATL, Wotherspoon ATL, Ansell RO, Brooks CJW (2000) Cholesterol oxidase: sources, physical properties and analytical applications. J Steroid Biochem Mol Biol 72(5):169–195

Mahley RW, Bersot TP (2017) Dyslipidemia and atherosclerosis. In: Kasper DL, Fauci AS, Hauser SL, Longo DL, Jameson JL, Loscalzo J (eds) Harrison's principles of internal medicine, 19th edn. McGraw-Hill Education, pp 2486–2500

Mendes MV, Recio E, Antón N, Guerra SM, Santos-Aberturas J, Martín JF, Aparicio JF (2007) Cholesterol oxidases act as signaling proteins for the biosynthesis of the polyene macrolide pimaricin. Chem Biol 14(3):279–290

Mohy-Eldin MS, El-Shamy AR, El-Sayed AB (2016) Recent advances in cholesterol oxidase immobilization: a review. Appl Biochem Biotechnol 180(3):503–519

Moradpour Z, Ghasemian A (2016) Protein engineering of microbial cholesterol oxidases: a molecular approach toward development of new enzymes with new properties. Appl Microbiol Biotechnol 100(10):4323–4336. https://doi.org/10.1007/s00253-016-7497-5. Epub 2016 Apr 11. PMID: 27063015

Purcell JP, Greenplate JT, Jennings JC, Ryerse JS, Pershing JC, Sims SR, Prange JC (1993) Cholesterol oxidase: a potent insecticidal protein against boll weevil larvae. Biochem Biophys Res Commun 196(3):1406–1413

Richmond W (1973) Preparation and properties of a cholesterol oxidase from Nocardia sp. and its application to the enzymatic assay of total cholesterol in serum. Clin Chem 19(12):1350–1356

Rose G, Shipley MJ (1980) Plasma lipids and mortality: a source of error. Lancet 1(8167):523–526. https://doi.org/10.1016/s0140-6736(80)92775-0. PMID: 6102243

Russell DW (2003) The enzymes, regulation, and genetics of bile acid synthesis. Annu Rev Biochem 72(1):137–174

Sasaki I, Gotoh H, Yamamoto R, Tanaka H, Takami K, Yamashita K, Yamashita J, Horio T (1982) Hydrophobic-ionic chromatography: its application to microbial glucose oxidase, hyaluronidase, cholesterol oxidase, and cholesterol esterase. J Biochem 91(5):1555–1561. https://doi.org/10.1093/oxfordjournals.jbchem.a133846. PMID: 6807971

Schatz A, Savard K, Pintner IJ (1949) J Bacteriol 53:117–125

Schmid A, Dordick JS, Hauer B, Kiener A, Wubbolts M, Witholt B (2001) Industrial biocatalysis today and tomorrow. Nature 409(6817):258–268

Shao M, Zhang X, Rao Z, Xu M, Yang T, Xu Z, Yang S (2019) Identification of steroid C27 monooxygenase isoenzymes involved in sterol catabolism and stepwise pathway engineering of Mycobacterium neoaurum for improved androst-1,4-diene-3,17-dione production. J Ind Microbiol Biotechnol 46(5):635–647. https://doi.org/10.1007/s10295-018-02135-5. Epub 2019 Feb 21. PMID: 30790119

Shapira M, Dobysh A, Liaudanskaya A, Aucharova H, Dzichenka Y, Bokuts V et al (2024) New insights into the substrate specificity of cholesterol oxidases for more aware application. Biochimie 220:1–10

Sharma M, Kango N, Sharma M (2018) Cholesterol oxidase from extremophilic microorganisms: a review. J Basic Microbiol 58(1):3–17

Srisawasdi P, Jearanaikoon P, Wetprasit N, Sriwanthana B, Kroll MH, Lolekha PH (2006) Application of Streptomyces and Brevibacterium cholesterol oxidase for total serum cholesterol assay by the enzymatic kinetic method. Clin Chim Acta 372(1–2):103–111

Stadtman TC, Cherkes A, Anfinsen CB (1954) Studies on the microbiological degradation of cholesterol. J Biol Chem 206:511–523

Smith AG, Brooks CJ (1976) Cholesterol oxidases: properties and applications. J Steroid 7(9):705–713

Suckling KE, Suckling CJ (1994) Cholesterol and membrane dynamics. Biol Chem Hoppe Seyler 375(1):1–13

Temani M, Baccar ZM, Mansour HB (2014) Activity of cholesterol oxidase immobilized on layered double hydroxide nanomaterials for biosensor application: Acacia salicina scavenging power of hypercholesterolemia therapy. Microelectron Eng 126:165–168

Thurnhofer H, Gains N, Mütsch B, Hauser H (1986) Cholesterol oxidase as a structural probe of biological membranes: its application to brush-border membrane. Biochim Biophys Acta 856(1):174–181

Torigoe S, Kuramoto K, Yoshioka T, Ohgiya S, Sugimori K (1999) Purification and characterization of a cholesterol oxidase from *Rhodococcus equi* with high activity towards water-insoluble substrates. Appl Microbiol Biotechnol 51(4):452–458

Turfitt GE (1947) Microbiological agencies in the degradation of steroids: II. Steroid utilization by the microflora of soils. J Bacteriol 54(5):557–562

Wang S, Chen S, Shang K, Gao X, Wang X (2021) Sensitive electrochemical detection of cholesterol using a portable paper sensor based on the synergistic effect of cholesterol oxidase and nanoporous gold. Int J Biol Macromol 189:356–362

Wei J, Li Y, Zhu L, Hu Y, Chen Z, Ma Y, Wei G (2012) Expression of cholesterol oxidase gene from *Streptomyces venezuelae* in tobacco plants confers resistance against cotton bollworm (*Helicoverpa armigera*). Appl Microbiol Biotechnol 93(5):1999–2007

Wei STS, Wu YW, Lee TH, Huang YS, Yang CY, Chen YL, Chiang YR (2018) Microbial functional responses to cholesterol catabolism in denitrifying sludge. mSystems 3(5):10–1128

Xin Y, Yang H, Xia X, Zhang L, Cheng C, Mou G et al (2011) Affinity purification of a cholesterol oxidase expressed in Escherichia coli. J Chromatogr B 879(13–14):853–858

Yamada H, Kondo A (1982) Cholesterol oxidase: purification, crystallization, and some properties. Methods Enzymol 89:516–522

Yao K, Wang FQ, Zhang HC, Wei DZ (2013) Identification and engineering of cholesterol oxidases involved in the initial step of sterols catabolism in Mycobacterium neoaurum. Metab Eng 15:75–87

Yazdi MT, Zahraei M, Aghaepour K, Kamranpour N (2001) Purification and partial characterization of a cholesterol oxidase from Streptomyces fradiae. Enzym Microb Technol 28(4–5):410–414

Yazdi MT, Yazdi ZT, Zarrini G, Ghasemian A, Sepehrizadeh Z, Olyaee NH (2008) Purification and characterization of extra-cellular cholesterol oxidase from *Rhodococcus sp.* PTCC 1633. Biotechnology (Faisalabad) 7(4):751–756. https://doi.org/10.3923/biotech.2008.751.756

Zaborowska-Mazurkiewicz M, Bizoń T, Matyszewska D, Fontaine P, Bilewicz R (2025) Oxidation of lipid membrane cholesterol by cholesterol oxidase and its effects on raft model membrane structure. Colloids Surf B: Biointerfaces 245:114191

Zhao S, Guo D, Zhu Q, Dou W, Guan W (2020) Display of microbial glucose dehydrogenase and cholesterol oxidase on the yeast cell surface for the detection of blood biochemical parameters. Biosensors (Basel) 11(1):13. https://doi.org/10.3390/bios11010013. PMID: 33396921; PMCID: PMC7823397

Enzyme Immobilization: Present Status and Scope for Pharmaceutical Application

11

Natasha Sharma Awasthi

Abstract

Enzymes are the biocatalysts which accelerate the biochemical reactions by lowering their activation energy and have exceptional specificity and efficiency, which makes them desirable for a wide array of industrial processes. However, challenges such as enzyme instability, limited reusability, and recovery inefficiencies have emphasized on the use of enzyme immobilization technologies. Immobilization involves restricting enzymes to inert, solid support materials, which enhances their stability, reusability, and tolerance to fluctuations in environmental conditions. This technique significantly reduces production costs and facilitates continuous processing. A variety of organic, inorganic, and nanomaterial carriers including alginate, chitosan, silica, and hybrid nanomaterials have been used to optimize enzyme performance. In pharmaceutical industries, enzyme immobilization plays a crucial role in the efficient synthesis of active pharmaceutical ingredients, therapeutic proteins, and optically pure compounds. Immobilized enzymes support high specificity and operational stability in key pharmaceutical reactions, including esterification, deracemization, and biotransformation. Techniques such as adsorption, covalent binding, cross-linking, and entrapment extend enzyme activity and process efficiency. Additionally, immobilized enzymes have therapeutic applications in treating metabolic, genetic, and infectious diseases. Commercial examples, such as Novozym 435, accentuate the industrial relevance of immobilized biocatalysts. This book chapter highlights on different methods of enzyme immobilization and the role of enzyme immobilization in progressing pharmaceutical biotechnology towards cost-effective, scalable, and environmentally sustainable biomedicine development.

N. S. Awasthi (✉)
University Centre for Research and Development, Chandigarh University, Mohali, Punjab, India
e-mail: natasha.e18536@cumail.in

© The Author(s), under exclusive license to Springer Nature Singapore Pte Ltd. 2026
V. Sharma, R. Salwan (eds.), *Pharmaceutically Relevant Microbial Enzymes*,
https://doi.org/10.1007/978-981-95-4941-2_11

Keywords

Enzymes · Immobilization · Carrier · Chemical reactions · Cost-effective · Reuse · Pharmaceutical

11.1 Introduction

Enzymes also known as biocatalysts are functional proteins which catalyze the biochemical reactions by lowering their activation energy. Certain enzymes such as ribozymes are nucleic acids by nature. Enzymes are present in both unicellular and multicellular organisms and are vital for proper growth and functioning of a living system (Kumar and Singh 2024). Enzymes are extremely selective in nature and can catalyze six different kinds of reactions, which are oxidation-reduction, group transfers within the molecules or between the molecules, hydrolysis, isomerization, ligation, bond cleavage, and bond formation reactions. The reactions catalyzed by the enzymes are highly specific, and each enzyme has its optimal set of temperature and pH conditions for its maximum efficiency (Nelson et al. 2008). Enzymes can remain viable extracellularly and perform array of reactions; these or their combinations are exploited in the industries to carry out number of processes; and chemical transformations of substrates to industrially important products. Enzymes have widespread application in chemical, food, pharmaceutical, textile, and waste management industries. The efficiency of the industrially important enzymes determines the cost and feasibility of the industrially important reactions (Liu et al. 2019). Microorganisms are the main source of these industrially important enzymes as their incorporation in industrial reactions is convenient and feasible and has potential for the scale-up due to their faster production. Also, the strain manipulation through various biotechnological approaches to manipulate the enzyme activity, over-expression, and specificity is also possible (Yang et al. 2017).

The cost of enzyme production, their recovery, and sensitivity to the environmental conditions affect the overall cost associated with the product development (Sóti et al. 2018). Traditional utilization of enzymes in different industries faced many challenges such as problem in the ease of handling, recovery and reuse of enzymes, and their sensitivity to the high concentrations of substrates and the products. To overcome these challenges in the industries, enzyme immobilization has been regarded as a huge success because it allows the smooth operation of both batch and flow bed methods of fermentation or reaction (Kujawa et al. 2021). Though microbial and enzyme immobilization has been used in the industries from the past four decades and has undergone so much progress, still the advantages these immobilized cells hold over the traditional fermentation techniques are numerous and are widely used and acceptable in the industries (Maghraby et al. 2023).

Enzyme activity can be increased several folds by their immobilization in the inert, non-reactive, and insoluble support materials because enzymes become more tolerant to high substrate and product concentrations and are able to withstand changes in the pH and temperature of the medium due to the protected environment.

The enzyme degradation reduces several folds and the final recovery and separation of enzyme from the final product becomes easy. Immobilized cells can be washed and reused. There are several methods of immobilization and several types of inert support materials in which immobilization can be carried out (Zhou and Chen 2001).

11.2 Definition and Advantages of Immobilization

The meaning of enzyme immobilization means restricting the enzymes in a certain physical space and rendering their mobility in a free medium. The catalytically active enzymes get physically confined to a particular place protected from the external environmental conditions to which an enzyme can be sensitive. Here substrate moves in, gets converted to the product, and moves out (Basso and Serban 2019). Restricting the gross movement of the enzyme is done through its attachment or enclosure in a suitable solid inert support material which is also known as carrier. This process separates enzymes in a discrete solid phase, which varies from the bulk phase containing substrates and products. Immobilization of an enzyme on insoluble matrix enables their reuse as the enzyme in matrix could be separated from the bulk and recovered using the simple technique of coarse filtration. Due to non-destructive separation of enzyme-carrier complexes, these can be washed again and reused; and it affects the overall cost of the reaction (Kosseva 2013). The primary goal of enzyme immobilization is to facilitate their prolonged use as catalysts. This technique also enhances catalyst stability by confining the enzymes within the porous structure of the support material (Kujawa et al. 2021).

Enzymes play a pivotal role in biotechnological applications, particularly hydrolytic enzymes, transferases, oxidoreductases, and lyases, which are integral to sustainable drug synthesis, food processing, biofuel production, and other industrial applications. There are several advantages to using the immobilized enzyme system over a free enzyme system. The immobilized enzyme system offers significant advantages over conventional methods, particularly in terms of lower production costs, higher productivity, shorter fermentation durations, and ease to control the reactions (DiCosimo et al. 2013; Eş et al. 2015). Firstly, a higher enzyme density or higher enzyme concentration per unit fermentation volume can be achieved than with conventional batch or continuous fermentation systems, resulting in a higher productivity in the biochemical reactors (Cao 2005). Secondly, using immobilized cells facilitates easy and efficient extraction of the product from the biocatalyst and the bulk (Ameur et al. 2014). Thirdly, enzyme immobilization enhances the prospect of recovery and reusability of enzymes. Fourthly, the danger of contamination reduces due to fast dilution rate in the continuously operated system (Kiyohara et al. 2003). Enzyme immobilization also enhances their stability by preserving their catalytic sites. In case of immobilized enzymes, beads could be reused without any loss of activity in continuous mode bioreactors, thereby reducing the cost of production to a great extent (Robescu and Bavaro 2025). Immobilized enzyme system offers the added benefit of easily converting batch processes to continuous operations, while also maintaining high enzyme densities, even at elevated dilution rates

(Ramanathan et al. 2000). Also, immobilized enzymes could act on a wide array of substrates. Furthermore, the bound enzyme systems exhibit significantly greater tolerance to fluctuations in reaction conditions such as temperature and pH and show reduced sensitivity to toxic substances in the liquid medium conditions under which soluble cells or enzymes might become inactive (Zhou and Chen 2001). Additionally, the products obtained are typically of high purity, making this technique particularly suitable for bioprocessing applications in the food and pharmaceutical industries (Basso and Serban 2019; Contesini et al. 2020). Since the benefits associated with the use of immobilized enzymes in chemical processes are many, industries have exploited them to their benefits. The advantages and disadvantages of immobilized cells over conventional fermentation methods are given in Fig. 11.1.

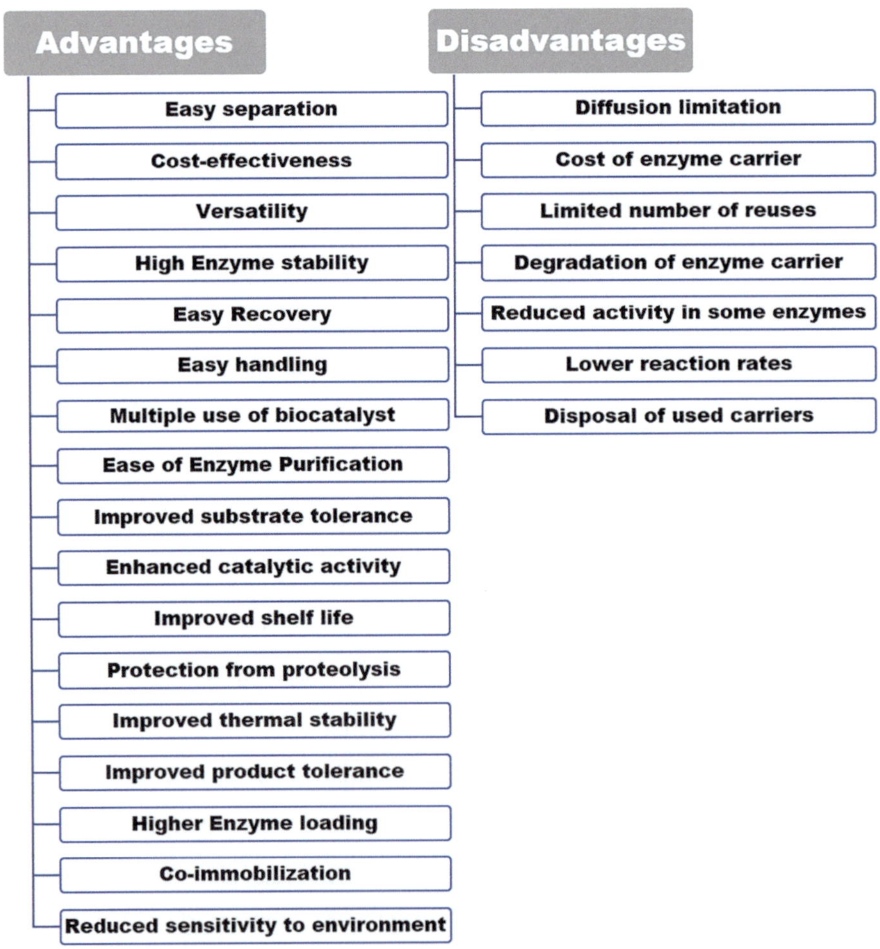

Fig. 11.1 Advantages and disadvantages of immobilized cells

11.3 Choice of a Support Material

Selection of a technique for immobilization and choice of a support material depends on factors such as the type of cells involved, the nature of the chemical transformation, and the characteristics of the substrate. In immobilization, cells are anchored to a distinct solid phase or carrier. This carrier, defined as the solid-phase material to which cells adhere, facilitates attachment through mechanisms such as adsorption, cross-linking, entrapment, or covalent bonding. An ideal support material must fulfill specific criteria to be suitable for the immobilization process. The support material should be inert, insoluble, cost-efficient and non-toxic (Vullo and Wachsman 2005). It should be capable of withstanding high temperatures, extreme pH levels, and elevated concentrations of substrates and products. Additionally, an important criterion for selecting a carrier is its resistance to microbial degradation. A good support material should facilitate the ease of diffusivity of substrates and the products and maintain the mildness of immobilization procedure (Robescu and Bavaro 2025).

The surface functional group, charge, pore size, particle size, and surface area of the carrier material are crucial for interaction with enzymes or their immobilization (Kujawa et al. 2021). There are various organic, inorganic, and nanomaterial carriers which can be used for enzyme immobilization. The most common examples of organic support materials are alginate, chitosan, carrageenan, agarose, activated carbon, cellulose, polyacrylamide, polypropylene membrane, and polyacrylonitrile. The inorganic support materials include silica, ceramic, alumina, clay, zirconia, and titania. Nanomaterials and hybrid nanomaterials have been extensively employed for enzyme immobilization. Commonly used materials include nanoporous alumina, silica nanoparticles, metal oxide nanoparticles, chitosan-based nanomaterials, and nanocellulose quantities. Hybrid nanomaterials such as gold nanocrystals, cellulose-coated magnetite nanoparticles, and zirconia-coated silica have also shown promise (Mohidem et al. 2023). Advancements in nanomaterials such as magnetic nanoparticles, silica-based carriers, and graphene oxide composites offer high surface area, facilitate easy enzyme recovery, and have significantly enhanced catalytic efficiency, thermal stability, reusability, and ease of downstream processing. These developments highlight the expanding role of immobilized enzymes in both industrial applications and therapeutic bioprocesses (Dandavate et al. 2009; Zlateski et al. 2014; El-Aziz et al. 2019). Recent research emphasizes the use of renewable, cost-effective, and readily available carrier materials in bulk such as lignocellulosic wastes (Girelli and Chiappini 2023).

11.4 Techniques of Immobilization

A variety of physical and chemical methods have been employed to immobilize enzymes or cells for accomplishing wide range of reactions in industries. These techniques have found widespread applications in fermentation processes; production of antibiotics and secondary metabolites; synthesis of organic acids and pharmaceuticals; as well as in food industry, medical and analytical fields; bioremediation; and the targeted treatment of pollutants (Vullo and Wachsman 2005). A number of immobilization techniques have been employed for enzyme immobilization using different support materials. The methods of immobilization can be broadly categorized into four types: entrapment, covalent binding, cross-linking, and adsorption. A combination of two or more of these techniques can also be utilized to enhance immobilization efficiency and stability (Guisan 2013). These methods have been described in detail below.

11.4.1 Adsorption

Adsorption involves physical binding of enzymes on the surface of inert carrier material. Enzyme immobilization through adsorption is based on electrostatic interactions, including Van der Waals forces, between the charged support surface and enzyme molecules (Liu and Dave 2022). Since it is a physical process of binding, it can be reversed by minor changes in pH, temperature, or ionic strength. For adsorption of enzymes, the properties of the support material are of utmost importance which include the composition of the carrier, surface charge, surface area, and the pore size. Adsorption equally depends upon the age, polarity, and charge on the enzyme (Brady and Jordaan 2009). Although operationally adsorption is economic and simple process to perform, the forces involved are the most complex. The bonds involved in adhesion can be weak ionic, hydrogen, co-ordinate, or a combination of any of these bonds, but the total interactions are mainly due to Van der Waals electrostatic forces. Adsorption has the advantage that enzyme activity hardly gets affected by the weak interactions between enzyme and support material (Shuai et al. 2017). The main drawback of this method is its sensitivity to changes in fermentation medium conditions such as ionic strength, pH, and stirring rates, which can significantly affect the extent of enzyme adsorption and lead to enzyme leakage with repeated use (Zivkovic et al. 2015).

11.4.2 Covalent Binding

Covalent binding is a chemical process, involves the formation of covalent bonds between activated inorganic support and enzymes, and is facilitated by a binding agent (Maghraby et al. 2023). Covalent binding of cells or enzymes to a solid matrix is a more stable method of immobilization and is not reversed by a change in pH or ionic strength. The covalent binding technique for enzyme immobilization is feasible because enzymes contain a variety of reactive groups such as hydroxyl, carboxyl, amino, amide, thiol, phenolic, ether, sulfydryl, and various substituted aromatic rings (Kujawa et al. 2021). In some cases, where carriers do not have reactive groups, modifiers having chemical groups are attached to the carrier to support their reaction with the enzymes. Covalent binding imparts stability to the enzyme-carrier complex, and it can thus withstand higher thermal fluctuations and wider ranges of pH (Shuai et al. 2017). But the major disadvantage of this method is that a low enzyme loading is achieved in comparison to the other techniques employed for enzyme immobilization. Other problem is that cells are exposed to potent reactive groups, which may lead to a loss in the structural integrity of an enzyme and might exert toxic effects on enzyme viability and sustainability. Also, this process requires long incubation periods to carry out immobilization (Liu and Dong 2020).

11.4.3 Cross-Linking

Cross-linking is a chemical method of immobilization where many enzyme molecules can bind together through cross-linking achieved through covalent binding between the carrier material and the enzyme molecules (Yamaguchi et al. 2018). Surface modification of the support molecule with different functional groups which are organic in nature is a general method used in cross-linking (Liu and Dave 2022). Enzymes are immobilized by cross-linking using reagents such as glutaraldehyde, aryl-haldes, and isothiocyanates. Among these, glutaraldehyde is extensively utilized agent due to its cost-effectiveness (Abd Rahman et al. 2025). Unlike covalent binding, this approach does not require a support material, as the crosslinker attaches two enzyme molecules through its two reactive groups. This reduces the dilution of enzyme and the overall cost of the reaction, unlike other methods of immobilization where non-catalytic support material accounts for more than 95% of the total mass of the immobilized enzyme reaction system (Sheldon 2011). However, the toxicity of the chemicals employed for cross-linking presents a strong limitation to the widespread application of this technique. It also produces low yield of the active enzymes due to inactivation by extensive cross-linking (Mateo et al. 2004).

11.4.4 Entrapment

Entrapment within polymer matrices is one of the most extensively studied and widely applied methods for enzyme immobilization (Ramanathan et al. 2000). This technique involves enclosing enzymes within a rigid polymer network that restricts their diffusion into the surrounding medium, while still allowing substrates to enter and products to exit. The core principle is to form a polymeric matrix around the enzymes or cells to be immobilized. The resulting gel must possess adequate porosity to facilitate the transport of reactants and products while effectively retaining the enzymes. This polymer network is typically inert and irreversibly formed (Kurzbaum et al. 2020). One of the advantages of this method is that it places minimal physical constraints on the enzymes, allowing them to remain mobile within their confined environment, while also shielding them from harsh external conditions (Ramakrishna and Prakasham 1999). A wide range of natural polymers, such as alginate, carrageenan, cellulose, agar, agarose, and gelatin as well as synthetic polymers like polyacrylamide and other acrylic derivatives, have been extensively studied and employed for enzyme entrapment. The gels can be formed under mild conditions

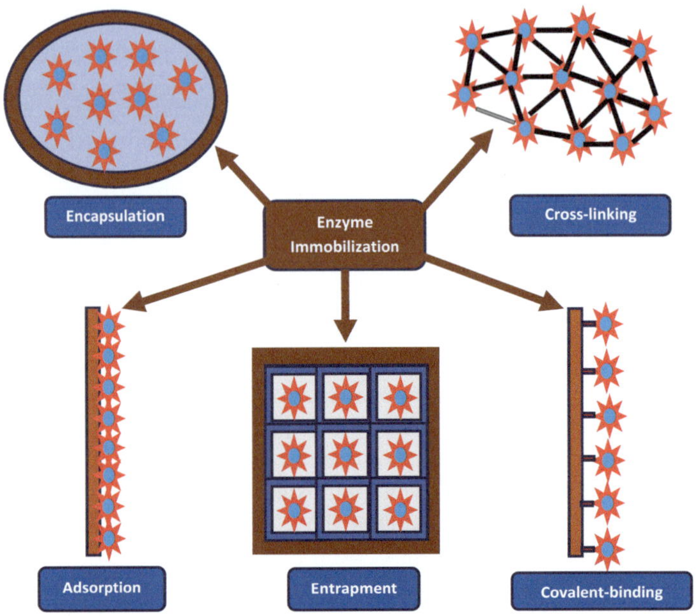

Fig. 11.2 Different methods of enzyme immobilization

and gels of different shapes and sizes with good porosity, and mechanical strength can be obtained at low costs (Ramdhan et al. 2020). The two most extensively studied and used carrier materials are natural hydrogels, and enzymes immobilized in them retain their catalytic activity for a longer duration, even after repeated use. Although the entrapment of enzymes in polymer gels is a popular and widely accepted method, it may not be the most effective, as it is subject to several limitations. Excessive enzyme leakage, limited O_2 supply in gel matrix, less diffusion of substrate through the solid support, and breakdown of gel beads caused by CO_2 generation during fermentation reactions are the major drawbacks of immobilization through entrapment (Dhanasekaran 2006; Rother and Nidetzky 2014) (Fig. 11.2).

11.5 Evaluation of an Immobilization Process

Immobilization process can be accessed through various parameters such as immobilization yield, immobilization efficiency, activity recovery, stability, selectivity, and reusability (Boudrant et al. 2020). Immobilization yield is calculated as the percentage of the enzyme activity retained on the support after immobilization. "Enzyme activity" is the amount of product formed per minute per gram or milligram of enzyme under specified conditions. "Immobilized activity" is determined by subtracting residual activity in the supernatant from the starting free enzyme activity. In some studies, immobilization yield is alternatively expressed based on the ratio of the amount of protein immobilized to the amount of protein added initially in the substrate (Robescu and Bavaro 2025).

$$\text{Immobilization yield}\left(\%\right) = \left(\text{Immobilized activity / Starting activity}\right) \times 100$$

"Immobilization efficiency" refers to the percentage of the immobilized enzyme that remains catalytically active, while "activity recovery" indicates the proportion of the original enzyme activity that is still functional after immobilization (Sheldon and van Pelt 2013).

$$\text{Immobilization efficiency}\left(\%\right) = \left(\text{Observed activity / Immobilized activity}\right) \times 100$$

$$\text{Activity recovery}\left(\%\right) = \left(\text{Observed activity / Starting activity}\right) \times 100$$

The major advantage of immobilization is enhanced stability, which is typically assessed by evaluating thermal and pH stability. "Thermal stability" involves incubating the enzyme at various temperatures and time intervals, while "pH stability" is examined by exposing the enzyme to buffers of different pH values at a constant temperature. "Residual activity" is measured in both cases to determine stability and is the ratio of enzyme activity determined after certain time of incubation to the enzyme activity measured at time zero. The selectivity of enzymes, including regioselectivity and chemoselectivity, is another important factor, as it can influence the structure and properties of the reaction products. "Reusability" is a crucial aspect for the commercial viability of immobilized enzymes, and it is ideally assessed by reusing the same enzyme in fresh reaction media across multiple cycles and tracking either relative activity or reaction progress such as hydrolysis or product concentration (Atiroglu et al. 2024).

Recently, new types of support materials for enzyme immobilization have been used such as chitosan and magnetic nanoparticles. To understand enzyme immobilization, structural characterization techniques are also employed. An increase in the distinctive absorption bands of amide, amino, or carboxyl groups originating from the enzymes can indicate successful immobilization through Fourier-transform infrared spectroscopy. Covalent attachment of an enzyme to the support may result in the appearance of additional absorption bands in the infrared spectrum. Scanning and transmission electron microscopy reveal morphological changes. For enzymes immobilized on nanoparticle supports, X-ray diffraction confirms crystal structures, and vibrating sample magnetometry assesses magnetic properties of nanoparticles. All these techniques provide a comprehensive understanding of the performance and structure of immobilized enzymes (Liu and Dave 2022).

11.6 Immobilized Cells in Pharmaceutical Industries

Pharmaceutical industrial biotechnology covers a broad spectrum of scientific and technological methods employed for the production of biopharmaceuticals or biologically derived medicinal products. This field primarily utilizes biotechnological processes, including the exploitation of genetically modified organisms and cultured cells, to obtain active pharmaceutical ingredients (APIs). The rapid progression in biotechnological innovations has profoundly influenced the landscape of drug discovery. Advancements in genomics, proteomics, and bioinformatics have facilitated the identification of novel biological targets, thereby expanding the scope of therapeutic development. Biopharmaceuticals, derived from complex biological sources such as human, animal, or microbial origins, include a diverse array of molecules ranging from sugars, proteins, and nucleic acids to vaccines, enzymes, living cells or tissues, blood products, immunosuppressants, hormones, cytokines, allergens, monoclonal antibodies, and recombinant DNA-based therapeutics (Ferreira et al. 2022). The tradition of utilizing fermentation in microbial cell cultures and enzymatic bioprocessing has long supported the production of biopharmaceuticals. The synthesis of such high-value compounds through traditional chemical methods poses substantial challenges; hence, the pharmaceutical industry has made considerable investments in enzyme immobilization technologies to develop enzyme catalysts capable of meeting stringent industrial demands. Enzymatic interventions have increasingly substituted conventional chemical methods in the synthesis of various active APIs (Contesini et al. 2020). Notable applications include the production of semi-synthetic antibiotics, resolution of racemic drug mixtures to obtain active enantiomers, and the synthesis of optically pure D- and L-amino acids. Enzymes have become essential tools in the pharmaceutical industry due to their versatility in mediating a wide range of reactions. These include hydrolysis, acylation and diacylation, deracemization, esterification, transesterification, and biotransformation reactions (Kumar and Singh 2024).

Enzyme immobilization has emerged as a key strategy in pharmaceutical manufacturing, enabling the synthesis of complex molecules with high specificity and purity. This technique has gained increasing attention due to its numerous advantages, including enhanced selectivity, operational safety, environmental sustainability, and the ability to produce high-yielding enantiomeric compounds and chemicals (Contesini et al. 2009). Techniques such as gel entrapment, cross-linking, adsorption, and covalent binding have been extensively applied to enhance the efficiency of fermentation processes by enhancing enzyme stability, reusability, and process efficiency. Immobilized enzymes facilitate prolonged operation by remaining within the bioreactor, thus addressing the limitations associated with free-enzyme fermentation (Rodrigues et al. 2021). The industrial-scale adoption of immobilized enzyme processes necessitates comprehensive economic assessments. Therefore, optimizing substrate-to-product conversion efficiency is critical. Innovative continuous fermentation techniques, such as enzyme immobilization, present significant cost-saving potential and offer a viable alternative to traditional batch fermentation methods. Immobilized enzyme systems are particularly advantageous in drug and therapeutic protein production due to their enhanced stability and reusability. Furthermore, the pharmaceutical industry has increasingly prioritized the incorporation of environmentally sustainable API manufacturing processes, with the aim of developing greener technologies that significantly reduce environmental impact (Degorska et al. 2023).

Table 11.1 summarizes various immobilization techniques used for enzyme immobilization to produce different pharmaceutically relevant products. Enzyme-based biocatalysis has increasingly replaced conventional chemical synthesis in pharmaceutical applications, particularly in the production of APIs and therapeutic agents such as gabapentinoids, rivastigmine, and antibiotics (Parmar et al. 2000; Lin et al. 2017; Su et al. 2024). In addition to these products, various vitamins and essential amino acids are also produced in pharmaceutical industries using immobilized enzymes. Beyond synthesis, enzymes have also demonstrated therapeutic potential. Enzyme therapies are employed in the treatment of numerous metabolic and genetic disorders, such as severe combined immunodeficiency, Gaucher's disease, Parkinson's disease, and phenylketonuria (Lwin 2004). Recent advances have extended enzyme applications to oncology and infectious disease management, particularly in cases where antibiotic resistance has rendered traditional treatments ineffective. Furthermore, enzymes such as proteases, lipases, and amylases serve as digestive aids, often administered individually or in combination to manage various gastrointestinal conditions (Ianiro et al. 2016). Novozym 435 (N435) is a commercially immobilized lipase widely used in research and industry. It contains *Candida antarctica* lipase B, expressed in *Aspergillus niger* and immobilized on Lewatit VP OC 1600, a hydrophobic, macroporous poly (methyl methacrylate) resin cross-linked with divinylbenzene (Ortiz et al. 2019).

11.7 Future Prospects of Enzymes in Pharmaceutical Industries

Immobilized enzymes offer significant advantages such as process efficiency by improving the performance of existing enzymes, environmental sustainability, and cost-effectiveness, compared to chemical synthesis, particularly in industrial applications. While the pharmaceutical sector faces unique challenges due to the need for enzymes to act on unnatural substrates under harsh conditions, advances in enzyme engineering and immobilization technologies are rapidly expanding their utility. Enzymes have already revolutionized API synthesis and therapeutic applications, and their role continues to grow in areas like metabolic disorders, cancer, and antibiotic resistance. With ongoing innovation, immobilized and engineered enzymes are poised to drive the next generation of sustainable pharmaceutical processes and novel therapies.

Table 11.1 Different pharmaceutically relevant drugs produced through enzyme immobilization

S.no.	Product	Target application	Immobilized enzyme	Type of immobilization	References
1.	Odanacatib	Osteoporosis treatment	Lipase B from *Candida antarctica*	Adsorption on octadecyl polymethacrylate resin	Truppo and Hughes (2011)
2.	Sofosbuvir	Treatment of hepatitis C	Lipase B from *Candida antarctica*	Adsorption on methacrylate/ divinylbenzene copolymer	Martin (2016)
3.	Aminopenicillanic acid	Antibiotic (Beta-lactams)	Penicillin G Amidase (hyd)	Covalent bonding on epoxy/amino methacrylate polymer	Parmar et al. (2000)
4.	Amoxicillin/ ampicillin	Antibiotic (Beta-lactams)	Penicillin G Amidase (syn)	Covalent on epoxy/ amino methacrylate polymer	Bruggink et al. (1998)
5.	Sitagliptin	Antidiabetic	Transaminase	Adsorption	Savile et al. (2010)
6.	GABA (Gamma-Aminobutyric Acid)	Analgesic, antianxiety, antidiabetic, and diuretic effects	Glutamate decarboxylase from *Lactobacillus fermentum* YS2	Enzyme adsorption on regenerated amorphous cellulose	Lin et al. (2017)
7.	GABA	Analgesic, antianxiety, antidiabetic, and diuretic effects	*Yarrowia lipolytica* glutamate decarboxylases	Electrospun nanofibrous membrane	Wu et al. (2024)
8.	Gabapentin	Treatment of epilepsy and neuropathic pain	Engineered nitrilase	Immobilized in diatomite cross-linked with polyethyleneimine and glutaraldehyde	Xue et al. (2017)
9.	Pregabalin	Postherpetic neuralgia and epilepsy	*Talaromyces thermophilus* Lipase	Epoxy resin	Ding et al. (2018)
10.	Ethyl (S)-3-cyano-5-methylhexanoate	A key chiral intermediate for Pregabalin	Lipase from *Thermomyces lanuginosus*	Crosslinked enzyme aggregates, CLEAs	Basetty and Kumaraguru (2022)
11.	S-rivastigmine	Alzheimer's and Parkinson's disease treatment	Amine transaminases from *Luminiphilus syltensis*, *Burkholderia multivorans*, *Vibrio fluvialis*	Cross-linked enzyme aggregates (CLEAs); glutaraldehyde-functionalized amine- and epoxy-beads	Heinks et al. (2023)

Table 11.1 Continued

S.no.	Product	Target application	Immobilized enzyme	Type of immobilization	References
12.	S-rivastigmine	Alzheimer's and Parkinson's disease treatment	Engineered alcohol dehydrogenase from *Lactobacillus brevis*	Immobilization on gold nano particles	Su et al. (2024)
14.	Diltiazem hydrochloride	Treatment of high blood pressure and chest pain	Lipase from *Serratia marcescens*	Amino-functionalized magnetic nanoparticles	Hu et al. (2009)
15.	Profens (2-aryl-propionicacids)	Non-steroidal anti-inflammatory drugs	Arylmalonate decarboxylase	Immobilization on an amino C2 acrylate carrier	Abmann et al. (2017)

11.8 Conclusions

Immobilized cells hold immense potential for the future of pharmaceutical and therapeutic applications, especially as enzymes continue to replace traditional chemical catalysts. The use of immobilized whole cells offers advantages such as enhanced enzyme stability, reusability, and ease of product separation. Immobilized enzymes can also play a crucial role in large-scale biotransformations, enabling more efficient, economical, and sustainable pharmaceutical manufacturing processes. With continued advancements in immobilization technologies and synthetic biology, immobilized enzyme systems are poised to become integral tools in the future of biopharmaceutical production and therapeutic enzyme delivery.

References

Abd Rahman NH, Rahman RA, Illias RM (2025) Investigating glutaraldehyde cross-linked starch as a hybrid support for immobilizing pectinase and xylanase for pectic-oligosaccharides production. Food Biosci 63:105713. https://doi.org/10.1016/j.fbio.2024.105713

Abmann M, Mügge C, Gaßmeyer SK, Enoki J, Hilterhaus L, Kourist R, Liese A, Kara S (2017) Improvement of the process stability of arylmalonate decarboxylase by immobilization for biocatalytic profen synthesis. Front Microbiol 8:448. https://doi.org/10.3389/fmicb.2017.00448

Ameur SB, Luminiţa Gîjiu C, Belleville M-P, Sanchez J, Paolucci-Jeanjean D (2014) Development of a multichannel monolith large-scale enzymatic membrane and application in an immobilized enzymatic membrane reactor. J Membr Sci 455:330–340. https://doi.org/10.1016/j.memsci.2013.12.026

Atiroglu V, Atiroğlu A, Atiroğlu A, Al-Hajri AS, Özacar M (2024) Green immobilization: enhancing enzyme stability and reusability on eco-friendly support. Food Chem 448:138978. https://doi.org/10.1016/j.foodchem.2024.138978

Basetty S, Kumaraguru T (2022) Preparation of enantiopure pregabalin intermediate using cross linked enzyme aggregates (CLEAs) in basket reactor. Biocatal Biotransformation 41(3):208–221. https://doi.org/10.1080/10242422.2021.2023507

Basso A, Serban S (2019) Industrial applications of immobilized enzymes—a review. Mol Catal 479:110607. Elsevier B.V. https://doi.org/10.1016/j.mcat.2019.110607

Boudrant J, Woodley JM, Fernandez-Lafuente R (2020) Parameters necessary to define an immobilized enzyme preparation. Process Biochem 90:66–80. https://doi.org/10.1016/j.procbio.2019.11.026

Brady D, Jordaan J (2009) Advances in enzyme immobilisation. Biotechnol Lett 31(11):1639–1650. https://doi.org/10.1007/s10529-009-0076-4

Bruggink A, Roos EC, de Vroom E (1998) Penicillin acylase in the industrial production of β-lactam antibiotics. Org Process Res Dev 2(2):128–133. https://doi.org/10.1021/op9700643

Cao L (2005) Immobilised enzymes: science or art? Curr Opin Chem Biol 9(2):217–226. https://doi.org/10.1016/j.cbpa.2005.02.014

Contesini FJ, da Silva VCF, Maciel RF, de Lima RJ, Barros FFC, de Oliveira Carvalho P (2009) Response surface analysis for the production of an enantioselective lipase from *Aspergillus niger* by solid-state fermentation. J Microbiol 47(5):563–571. https://doi.org/10.1007/s12275-008-0279-8

Contesini FJ, Davanço MG, Borin GP, Vanegas KG, Cirino JPG, de Melo RR, Mortensen UH, Hildén K, Campos DR, Carvalho PDO (2020) Advances in recombinant lipases: production, engineering, immobilization and application in the pharmaceutical industry. Catalysts 10(9):1–33. https://doi.org/10.3390/catal10091032

Dandavate V, Keharia H, Madamwar D (2009) Ethyl isovalerate synthesis using *Candida rugosa* lipase immobilized on silica nanoparticles prepared in nonionic reverse micelles. Process Biochem 44(3):349–352. https://doi.org/10.1016/j.procbio.2008.11.001

Degorska O, Szada D, Jesionowski T, Zdarta J (2023) A biocatalytic approach for resolution of 3-hydroxy-3-phenylpropanonitrile with the use of immobilized enzymes stabilized with ionic liquids. Comput Struct Biotechnol J 21:1593–1597. https://doi.org/10.1016/j.csbj.2023.02.026

Dhanasekaran D (2006) Studies on free and immobilised cells of *Bacillus* species on the production of alpha-amylase. Int J Microbiol 2(2):2039–2047

DiCosimo R, McAuliffe J, Poulose AJ, Bohlmann G (2013) Industrial use of immobilized enzymes. Chem Soc Rev 42(15):6437. https://doi.org/10.1039/c3cs35506c

Ding X, Tang X-L, Zheng R-C, Zheng Y-G (2018) Efficient biocatalytic synthesis of chiral intermediate of pregabalin using immobilized *Talaromyces thermophilus* lipase. Biomed Res Int 2018:1–7. https://doi.org/10.1155/2018/6192059

El-Aziz AMA, Shaker MA, Shaaban MI (2019) Enhanced biocatalytic activity of recombinant lipase immobilized on gold nanoparticles. Curr Pharm Biotechnol 20(6):497–505. https://doi.org/10.2174/1389201020666190416144650

Eş I, Vieira JDG, Amaral AC (2015) Principles, techniques, and applications of biocatalyst immobilization for industrial application. Appl Microbiol Biotechnol 99(5):2065–2082. https://doi.org/10.1007/s00253-015-6390-y

Ferreira PG, de Carvalho AS, Santos WC, Forezi LSM, da Silva FC, Ferreira VF (2022) Sustainable synthetic strategies for the preparation of low molecular weight drugs by biotech routes. J Braz Chem Soc 33(7):664–680). Sociedade Brasileira de Quimica. https://doi.org/10.21577/0103-5053.20220044

Girelli AM, Chiappini V (2023) Renewable, sustainable, and natural lignocellulosic carriers for lipase immobilization: a review. J Biotechnol 365:29–47. https://doi.org/10.1016/j.jbiotec.2023.02.003

Guisan JM (2013) New opportunities for immobilization of enzymes. In: Immobilization of enzymes and cells, Methods in molecular biology. Humana Press, pp 1–13. https://doi.org/10.1007/978-1-62703-550-7_1

Heinks T, Montua N, Teune M, Liedtke J, Höhne M, Bornscheuer UT, Fischer von Mollard G (2023) Comparison of four immobilization methods for different transaminases. Catalysts 13(2):300. https://doi.org/10.3390/catal13020300

Hu B, Pan J, Yu H-L, Liu J-W, Xu J-H (2009) Immobilization of *Serratia marcescens* lipase onto amino-functionalized magnetic nanoparticles for repeated use in enzymatic synthesis of Diltiazem intermediate. Process Biochem 44(9):1019–1024. https://doi.org/10.1016/j.procbio.2009.05.001

Ianiro G, Pecere S, Giorgio V, Gasbarrini A, Cammarota G (2016) Digestive enzyme supplementation in gastrointestinal diseases. Curr Drug Metab 17(2):187–193. https://doi.org/10.2174/138920021702160114150137

Kiyohara PK, Lima UDA, Santos HDS, Santos PDS (2003) Comparative study between yeasts immobilized on alumina beads and on membranes prepared by two routes. Braz J Microbiol 34:129–137. https://doi.org/10.1590/S1517-83822003000200008

Kosseva MR (2013) Use of immobilized biocatalyst for valorization of whey lactose. In: Food industry wastes. Elsevier, pp 137–156. https://doi.org/10.1016/B978-0-12-391921-2.00008-1

Kujawa J, Głodek M, Li G, Al-Gharabli S, Knozowska K, Kujawski W (2021) Highly effective enzymes immobilization on ceramics: requirements for supports and enzymes. Sci Total Environ 801:149647. Elsevier B.V. https://doi.org/10.1016/j.scitotenv.2021.149647

Kumar P, Singh J (2024) Exploring enzyme biotechnology's role in green chemistry and advancements in pharmaceutical technologies. In: Enzyme biotechnology for environmental sustainability. Academic Press, pp 465–495. https://doi.org/10.1016/B978-0-443-22072-2.00015-2

Kurzbaum E, Raizner Y, Kuc ME, Kulikov A, Hakimi B, Kruh LI, Menashe O (2020) Phenol biodegradation by bacterial cultures encapsulated in 3D microfiltration-membrane capsules. Environ Technol 41(22):2875–2883. https://doi.org/10.1080/09593330.2019.1587005

Lin Q, Li D, Qin H (2017) Molecular cloning, expression, and immobilization of glutamate decarboxylase from *Lactobacillus fermentum* YS2. Electron J Biotechnol 27:8–13. https://doi.org/10.1016/j.ejbt.2017.03.002

Liu Y, Dave D (2022) Recent progress on immobilization technology in enzymatic conversion of marine by-products to concentrated omega-3 fatty acids. Green Chem 24(3):1049–1066). Royal Society of Chemistry. https://doi.org/10.1039/d1gc03127a

Liu DM, Dong C (2020) Recent advances in nano-carrier immobilized enzymes and their applications. Process Biochem 92:464–475. https://doi.org/10.1016/j.procbio.2020.02.005

Liu Q, Xun G, Feng Y (2019) The state-of-the-art strategies of protein engineering for enzyme stabilization. Biotechnol Adv 37(4):530–537. https://doi.org/10.1016/j.biotechadv.2018.10.011

Lwin A (2004) Glucocerebrosidase mutations in subjects with parkinsonism. Mol Genet Metab 81(1):70–73. https://doi.org/10.1016/j.ymgme.2003.11.004

Maghraby YR, El-Shabasy RM, Ibrahim AH, Azzazy HMES (2023) Enzyme immobilization technologies and industrial applications. ACS Omega 8(6):5184–5196). American Chemical Society. https://doi.org/10.1021/acsomega.2c07560

Martin N (2016) A process for preparing a crystalline form of sofosbuvir (WO 2016/156512 A)

Mateo C, Palomo JM, van Langen LM, van Rantwijk F, Sheldon RA (2004) A new, mild cross-linking methodology to prepare cross-linked enzyme aggregates. Biotechnol Bioeng 86:273–276. https://doi.org/10.1002/bit.20033

Mohidem NA, Mohamad M, Rashid MU, Norizan MN, Hamzah F, Mat HB (2023) Recent advances in enzyme immobilisation strategies: an overview of techniques and composite carriers. J Compos Sci 7(12):488. https://doi.org/10.3390/jcs7120488

Nelson DL, Lehninger AL, Cox MM (2008) Lehninger principles of biochemistry, 5th edn. W.H. Freeman

Ortiz C, Ferreira ML, Barbosa O, dos Santos JCS, Rodrigues RC, Berenguer-Murcia Á, Briand LE, Fernandez-Lafuente R (2019) Novozym 435: the "perfect" lipase immobilized biocatalyst? Cat Sci Technol 9(10):2380–2420. https://doi.org/10.1039/C9CY00415G

Parmar A, Kumar H, Marwaha SS, Kennedy JF (2000) Advances in enzymatic transformation of penicillins to 6-aminopenicillanic acid (6-APA). Biotechnol Adv 18(4):289–301. https://doi.org/10.1016/S0734-9750(00)00039-2

Ramakrishna SV, Prakasham RS (1999) Microbial fermentations with immobilized cells. Curr Sci 77:87–100

Ramanathan K, Pandey SS, Kumar R, Gulati A, Murthy ASN, Malhotra BD (2000) Covalent immobilization of glucose oxidase to poly (O-amino benzoic acid) for application to glucose biosensor. J Appl Polym Sci 78(3):662–667. https://doi.org/10.1002/1097-4628(20001017)78:3<662::AID-APP220>3.0.CO;2-T

Ramdhan T, Ching SH, Prakash S, Bhandari B (2020) Physical and mechanical properties of alginate based composite gels. Trends Food Sci Technol 106:150–159. https://doi.org/10.1016/j.tifs.2020.10.002

Robescu MS, Bavaro T (2025) A comprehensive guide to enzyme immobilization: all you need to know. Molecules 30(4):939. https://doi.org/10.3390/molecules30040939

Rodrigues RC, Berenguer-Murcia Á, Carballares D, Morellon-Sterling R, Fernandez-Lafuente R (2021) Stabilization of enzymes via immobilization: multipoint covalent attachment and other stabilization strategies. Biotechnol Adv 52:107821. https://doi.org/10.1016/j.biotechadv.2021.107821

Rother C, Nidetzky B (2014) Enzyme immobilization by microencapsulation: methods, materials, and technological applications. In: Encyclopedia of industrial biotechnology. Wiley, pp 1–21. https://doi.org/10.1002/9780470054581.eib275

Savile CK, Janey JM, Mundorff EC, Moore JC, Tam S, Jarvis WR, Colbeck JC, Krebber A, Fleitz FJ, Brands J, Devine PN, Huisman GW, Hughes GJ (2010) Biocatalytic asymmetric synthesis of chiral amines from ketones applied to sitagliptin manufacture. Science 329(5989):305–309. https://doi.org/10.1126/science.1188934

Sheldon RA (2011) Characteristic features and biotechnological applications of cross-linked enzyme aggregates (CLEAs). Appl Microbiol Biotechnol 92(3):467–477. https://doi.org/10.1007/s00253-011-3554-2

Sheldon RA, van Pelt S (2013) Enzyme immobilisation in biocatalysis: why, what and how. Chem Soc Rev 42(15):6223–6235. https://doi.org/10.1039/C3CS60075K

Shuai W, Das RK, Naghdi M, Brar SK, Verma M (2017) A review on the important aspects of lipase immobilization on nanomaterials. Biotechnol Appl Biochem 64(4):496–508. https://doi.org/10.1002/bab.1515

Sóti V, Lenaerts S, Cornet I (2018) Of enzyme use in cost-effective high solid simultaneous saccharification and fermentation processes. J Biotechnol 270:70–76. https://doi.org/10.1016/j.jbiotec.2018.01.020

Su G, Ran L, Liu C, Qin Z, Teng H, Wu S (2024) Directed evolution and immobilization of *Lactobacillus brevis* alcohol dehydrogenase for chemo-enzymatic synthesis of Rivastigmine. Chem Eur J 30(32):e202400454. https://doi.org/10.1002/chem.202400454

Truppo MD, Hughes G (2011) Development of an improved immobilized CAL-B for the enzymatic resolution of a key intermediate to Odanacatib. Org Process Res Dev 15(5):1033–1035. https://doi.org/10.1021/op200157c

Vullo DL, Wachsman MB (2005) A simple laboratory exercise for ethanol production by immobilized bakery yeasts (*Saccharomyces cerevisiae*). J Food Sci Educ 4(3):53–55. https://doi.org/10.1111/j.1541-4329.2005.tb00060.x

Wu LT, Huang YH, Hsieh LS (2024) Production of γ-aminobutyric acid by immobilization of two *Yarrowia lipolytica* glutamate decarboxylases on electrospun nanofibrous membrane. Int J Biol Macromol 278:135046. https://doi.org/10.1016/j.ijbiomac.2024.135046

Xue YP, Zhong HJ, Zou SP, Zheng YG (2017) Efficient chemoenzymatic synthesis of gabapentin by control of immobilized biocatalyst activity in a stirred bioreactor. Biochem Eng J 125:190–195. https://doi.org/10.1016/j.bej.2017.06.008

Yamaguchi H, Kiyota Y, Miyazaki M (2018) Techniques for preparation of cross-linked enzyme aggregates and their applications in bioconversions. Catalysts 8(5):174. https://doi.org/10.3390/catal8050174

Yang H, Li J, Du G, Liu L (2017) Microbial production and molecular engineering of industrial enzymes: challenges and strategies. In: Biotechnology of microbial enzymes. Academic Press, pp 151–165. https://doi.org/10.1016/B978-0-12-803725-6.00006-6

Zhou QZK, Chen XD (2001) Effects of temperature and pH on the catalytic activity of the immobilized β-galactosidase from Kluyveromyces lactis. Biochem Eng J 9(1):33–40. https://doi.org/10.1016/S1369-703X(01)00118-8

Zivkovic LTI, Živković LS, Babić BM, Kokunešoski MJ, Jokić BM, Karadžić IM (2015) Immobilization of *Candida rugosa* lipase by adsorption onto biosafe meso/macroporous silica and zirconia. Biochem Eng J 93:73–83. https://doi.org/10.1016/j.bej.2014.09.012

Zlateski V, Fuhrer R, Koehler FM, Wharry S, Zeltner M, Stark WJ, Moody TS, Grass RN (2014) Efficient magnetic recycling of covalently attached enzymes on carbon-coated metallic nanomagnets. Bioconjug Chem 25(4):677–684. https://doi.org/10.1021/bc400476y

MIX
Papier aus verantwortungsvollen Quellen
Paper from responsible sources
FSC® C105338

If you have any concerns about our products,
you can contact us on
ProductSafety@springernature.com

In case Publisher is established outside the EU,
the EU authorized representative is:
Springer Nature Customer Service Center GmbH
Europaplatz 3, 69115 Heidelberg, Germany

Printed by Libri Plureos GmbH
in Hamburg, Germany